AF790916

Graphene Nanoribbons

Graphene Nanoribbons

Edited by
Luis Brey
Instituto de Ciencia de Materiales de Madrid-CSIC, Madrid, Spain

Pierre Seneor
Unité Mixte de Physique CNRS/Thales, Paris, France

Antonio Tejeda
Université Paris Sud, Paris, France

IOP Publishing, Bristol, UK

© IOP Publishing Ltd 2020

All rights reserved. No part of this publication may be reproduced, stored in a retrieval system or transmitted in any form or by any means, electronic, mechanical, photocopying, recording or otherwise, without the prior permission of the publisher, or as expressly permitted by law or under terms agreed with the appropriate rights organization. Multiple copying is permitted in accordance with the terms of licences issued by the Copyright Licensing Agency, the Copyright Clearance Centre and other reproduction rights organizations.

Permission to make use of IOP Publishing content other than as set out above may be sought at permissions@ioppublishing.org.

Luis Brey, Pierre Seneor and Antonio Tejeda have asserted their right to be identified as the authors of this work in accordance with sections 77 and 78 of the Copyright, Designs and Patents Act 1988.

ISBN 978-0-7503-1701-6 (ebook)
ISBN 978-0-7503-1699-6 (print)
ISBN 978-0-7503-1700-9 (mobi)

DOI 10.1088/978-0-7503-1701-6

Version: 20191101

IOP ebooks

British Library Cataloguing-in-Publication Data: A catalogue record for this book is available from the British Library.

Published by IOP Publishing, wholly owned by The Institute of Physics, London

IOP Publishing, Temple Circus, Temple Way, Bristol, BS1 6HG, UK

US Office: IOP Publishing, Inc., 190 North Independence Mall West, Suite 601, Philadelphia, PA 19106, USA

Contents

Preface

Graphene, a one-atom thick layer of carbon atoms organized following a honeycomb lattice has been at the forefront of research for more than a decade. This is thanks to an uncommon combination of robustness and exciting novel electronic properties. These electronic properties of graphene become even more exceptional when one additional dimension is pushed down to the nanometer scale. These objects, called graphene nanoribbons, develop new properties significantly different from their graphene counterpart that deserve particular attention and require sophisticated tools, or miniaturized transistors and circuits, to be observed. In this book, we compile in seven chapters the fundamentals of graphene nanoribbons and some of the current research lines, both from theoretical and experimental points of view. We pay particular attention to the production methods using either physical or chemical approaches.

The first part of the book deals with production methods of graphene nanoribbons. The first chapter by Anna Miettinen *et al*, describes a top-down approach where a lithography template is used to grow graphene nanoribbons on technologically relevant SiC substrates. The authors also study the influence of the nanoribbon edge and the substrate polytype on the properties of the ribbons. The second chapter by Joffrey Pigeat *et al* describes other production methods of graphene nanoribbons, this time based on chemical synthesis, using a bottom-up approach.

The second part of the book deepens into the properties of graphene nanoribbons. Chapter 3 by Pilar Sánchez López *et al* initiates this part by discussing the dissipationless spin current along the electronic states located at the edges of graphene nanoribbons. J L Lado *et al* continue the discussion of the electronic states of graphene nanoribbons in chapter 4, focusing on non-trivial electronic states, such as those appearing in topological phases or quantum spin liquids. They also analyze the magnetic ordering at the edges of the ribbons and the superconductivity induced by proximity effect. This is followed by two chapters dealing with electronic transport from the theoretical and experimental point of view. Chapter 5, by Leonor Chico *et al* reviews the transport properties of metallic ribbons, and analyzes their role as valley filters and their applications in spintronics. Chapter 6, by Christoph Tegenkamp *et al* focuses on the transport of graphene nanoribbons on SiC using a four-tip STM to study the transport at the nanometer scale. The ballistic transport along these ribbons as well as the Klein tunneling at p-n junctions are described. The final chapter, by Stephan Roche *et al* analyzes the role of the inhomogeneities in the nanoribbons, either due to randomly attached functional groups, doping impurities, structural defects or magnetic fields. The impurity resonances, the transport mobility gaps and finally the transistor functionalities are presented.

Editor biographies

Luis Brey

Luis Brey is a research professor at the Instituto de Ciencia de Materiales de Madrid of the Consejo Superior de Investigaciones Científicas. He holds a PhD (1987) in Physics from the Universidad Autónoma de Madrid. He spent two years as a post-doctoral researcher at the Physics Department at Harvard University, and he has been a research visitor at Cambridge University, Max Planck Institute and Bell Labs at Murray Hill. He is a condensed matter theorist. At present, his main interests are in the electronic, magnetic and optical properties of two-dimensional materials.

Pierre Seneor

Pierre Seneor has been a Professor at Université Paris-Sud (now Université Paris-Saclay), Orsay, since 2003 and became junior member of the Institut Universitaire de France in 2010. He completed his PhD from Ecole Polytechnique in 2000 under the supervision of A Fert (Nobel Prize 2007). He then worked as a post-doc researcher at the California Institute of Technology (Caltech) before coming back to France to the CNRS/Thales joint physics laboratory where he is now. His interests focus on nanophysics, more specifically spintronics from single electron devices, to molecular spintronics and more recently 2D materials.

Antonio Tejeda

Antonio Tejeda works for CNRS, the main French Research Center since 2004, where he is now a Research Professor. He currently works at the Laboratoire de Physique des Solides located at the University of South Paris and is associated scientist to SOLEIL synchrotron. He received his PhD at Autonoma University in Madrid in 2003 and he worked as a post-doc researcher at Denis Diderot University in Paris under a Marie Curie Fellowship. His area of expertise are low dimensional systems, often combining structural techniques and electronic spectroscopy.

Contributors

Johannes Aprojanz
Leibniz Universitaet Hannover, Germany

Jens Baringhaus
Leibniz Universitaet Hannover, Germany

Blanca Biel
Universidad de Granada, Spain

Stéphane Campidelli
CEA, France

Jean-Cristophe Charlier
Institute of Condensed Matter and Nanosciences, Belgium

Leonor Chico
CSIC, Spain

Edward H Conrad
Georgiatech, United States of America

Alessandro Cresti
Univ. Grenoble Alpes, Univ. Savoie Mont Blanc, France

Joaquín Fernández-Rossier
International Iberian Nanotechnology Laboratory, Portugal

Jhon W González
Universidad Técnica Federico Santa María, Chile

José Luis Lado
Aalto University, Finland

Jean-Sébastien Lauret
ENS Paris Saclay, France

M Pilar López Sancho
CSIC, Spain

Alejandro López-Bezanilla
Los Alamos National Laboratory, United States of America

Anna Miettinen
Georgiatech, United States of America

M Carmen Muñoz
CSIC, Spain

Meredith S Nevius
Exponent, United States of America

Ricardo Ortiz
Universidad de Alicante, Spain

Marta Pelc
Donostia International Physics Center, Spain

Joffrey Pijeat
CEA, France

Stephan Roche
Catalan Institute of Nanoscience and Nanotechnology, Spain
Institució Catalana de Recerca i Estudis Avancats, Spain

Hernán Santos
Universidad Autónoma de Madrid, Spain

Christoph Tegenkamp
Technische Universität Chemnitz, Germany

Symbols

α Temperature coefficient of linear expansion (K^{-1})β
Temperature coefficient of volume expansion (K^{-1})γ
Ratio of heat capacities
ε Permittivity
κ Dielectric constant
λ Wavelength (m)
ρ Density (kg m^{-3})
B Magnetic field (T)
C Molar heat capacity (J $kg^{-1}K^{-1}$)*f*
Frequency
k Thermal conductivity ($Wm^{-1}K^{-1}$)*R*
Ideal gas constant (8.31 $Jmol^{-1}K^{-1}$)

IOP Publishing

Graphene Nanoribbons

Luis Brey, Pierre Seneor and Antonio Tejeda

Chapter 1

The growth and structure of epitaxial graphene nanoribbons

A Miettinen, M S Nevius and E H Conrad

Epitaxial graphene (EG) is grown from SiC(0001). EG is orientationally ordered with the SiC substrate and its thickness can be controlled down to single layers. These properties have been used to grow narrow graphene ribbons with predefined edge types from steps lithographically pre-patterned in the SiC substrate. Controlling the growth of graphene on the step walls has significantly advanced in the last decade. Optimized growth recipes have enabled numerous techniques to study the structure of these 'sidewall' graphene ribbons and correlate the ribbon's structure to its electronic properties. In this review, we cover the state of the art EG ribbon growth and show how the electronic properties of these ribbons are related to edge type and the substrate SiC polytype.

1.1 Introduction

The earliest motivation for graphene research was to use this 2D material for novel electronics [1, 2]. One avenue to graphene electronics is to make graphene semiconduct by opening a predicted finite size effect (or confinement) band gap in narrow graphene ribbons [3, 4]. Producing a technologically relevant band gap (~0.5 – 1 eV), however, not only requires ribbons to be only a few nanometers wide, but also requires that the ribbon edges have a specific orientation [4–8]. Graphene's two primary edge types are armchair (AC) and zigzag (ZZ). They run perpendicular (AC-edges) or parallel (ZZ-edges) to the graphene $\langle 01 \rangle_g$ direction (see figure 1.1(a)). Early tight binding (TB) calculations show that ZZ-edges will be metallic [3, 4]. The metallic state is associated with electrons localized to the ribbon edges. At the same time, TB calculations showed that a subset of AC-edge ribbons are metallic while the others are semiconducting with a gap inversely proportional to their width [3, 4]. Later, *ab initio* calculations showed that both ZZ- and all AC-edges ribbons would be semiconducting [5, 6]. In the case of ZZ-edge ribbons, the

doi:10.1088/978-0-7503-1701-6ch1 © IOP Publishing Ltd 2020

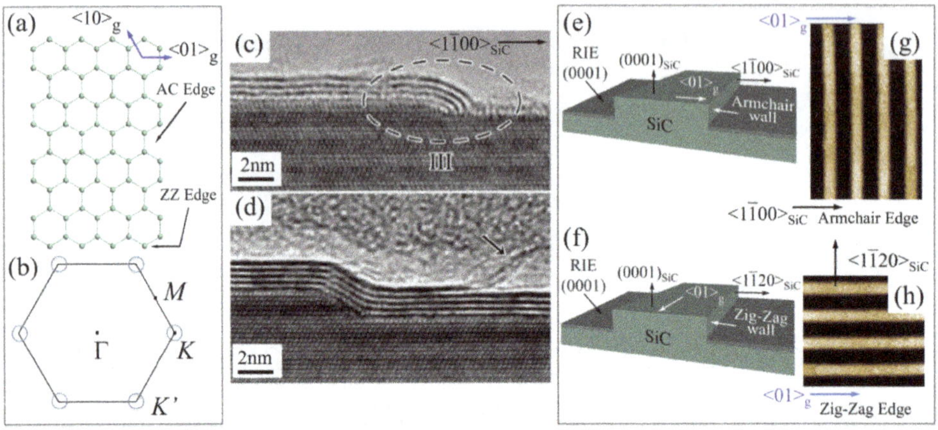

Figure 1.1. (a) The graphene lattice showing AC- and ZZ-edges. (b) The graphene Brillouin zone orientation relative to the graphene lattice in (a). (c) and (d) HRTEM images showing two different growth modes of multilayer graphene growing over natural steps on SiC(0001). The SiC steps are perpendicular to the $\langle 1\bar{1}00\rangle_{SiC}$ direction (i.e. graphene's AC-edges running into the plane of the figure). Graphene layers are observed as dark lines. Reprinted from [9], Copyright (2010), with permission from Elsevier. (e), (f) Schematics of etched plateaus in SiC whose edges are aligned with (e) AC- and (f) ZZ-graphene ribbons, respectively. AFM images of graphitized SiC plateaus with (g) AC-edge graphene and (h) ZZ-edge graphene. Dark areas are the trench bottoms.

semiconducting gap is due to a splitting of the edge state due to electron–electron interactions breaking the edge state degeneracy.

Demonstrating that a finite size gap could be opened proved daunting. The earliest attempt, using lithographically patterned ribbons, indicated that a gap did open [10]. However, later experiments [11–13] and theoretical calculations [8, 14] showed that 'necks' in the ribbons due to edge disorder produce charge puddles that dominate transport. The Coulomb blockade problem at these necks led to a 'transport gap' that ultimately dominates and thus degrades the ideal ribbon's transport properties. In other words, not only must the lithography be capable of producing sufficiently narrow graphene ribbons, it must also achieve an exceptionally high degree of edge order if graphene ribbon devices are to be developed. Current lithography simply cannot achieve these limits.

Many clever workarounds were investigated to solve the patterning problem. Carbon nanotubes have been unzipped using an Ar plasma to give widths as low as ~7 nm [15], and chemical vapor deposition (CVD) grown graphene has been cut to get widths down to 3.5 nm using a STM to achieve well ordered edges [16]. While impressive, both techniques are simply not scalable, making it impossible to use them in large scale circuits. Graphene ribbons, both AC- and ZZ-edged, have also been fabricated from chemical precursors on metal surfaces [17–22]. Band gaps larger than 2.5 eV have been directly measured on AC-ribbons fabricated on vicinal Au surfaces [19]. While these chemical synthesis methods of graphene ribbons produce well defined edge types and controllable widths, they also have fabrication issues. These ribbons are typically only 20–40 nm long, and the fabrication process is still limited to metal substrates. These limitations make a transfer step necessary, thus saddling these ribbons with the same scalability and circuit integration problems as CVD and nanotube ribbons.

In this review, we focus on the only known scalable approach to fabricating graphene ribbons; EG nano-ribbons grown on SiC [23]. These ribbons can be grown not only in predetermined locations but can also be grown with predetermined and highly ordered edge types. Since transport studies of these ribbons will be covered separately in this collection, we will focus on the current state of their fabrication, structure, and what is known about their electronic properties. We will show that EG ribbon growth depends on Si vapor pressure, temperature, time, and SiC polytype. We will also show that their structure is more complicated than early assumptions made about these ribbons. Rather than being ideally terminated and noninteracting with the SiC substrate, these ribbons have asymmetric terminations and grow on complicated SiC facet structures that can locally cause the ribbons to bond to the SiC. Finally, their electronic and structural properties have now been measured and lead to surprises that require changing our interpretation of early transport measurements. This work is intended to convey what is known about these ribbons and what remains to be learned before they can be reproducibly integrated into an all carbon electronic system.

1.2 Epitaxial ribbon production

While the bulk of this review will discuss patterned EG sidewall ribbons, there have been a some interesting works on stepped surfaces using off axis cut SiC. In particular, we will briefly review the work of Kajiwara *et al* [24] in section 1.2.2. They show a way to grow ribbons on narrow nano-(0001) terraces rather than on the sidewall facet. Graphene growth on patterned SiC will be discussed in sections 1.2.1 and 1.2.3.

1.2.1 EG ribbon methods and geometry

Graphene grows epitaxially on the SiC(0001) (Si-face) surface [25, 26]. Because EG is epitaxial with SiC(0001), EG's orientation after growth is well defined and set by the known SiC lattice orientation. Specifically, EG grows rotated 30° relative to the SiC $\langle 10\bar{1}0 \rangle$ direction [26]. Since growth on the SiC(000$\bar{1}$) C-face is both rotationally complicated and leads to multiple layer films [26], sidewall ribbon growth in this review is limited to the Si-face. We begin in section 1.2.1 with a general discussion of the approach for producing EG sidewall ribbons with known edge types. We then discuss the pre-growth processing necessary to make trenches in SiC. EG-ribbons grow on the walls of these trenches. Since AC- and ZZ-edge ribbons grow differently on SiC, we will treat their growth separately.

The idea of patterning trenches in SiC(0001) and then growing graphene on their sidewalls, thus leading to predetermined shapes and edge types, originated with the work of de Heer *et al* [23, 27, 28]. It was known from high resolution transmission electron microscopy (HRTEM) work that graphene growth on the Si-face of SiC begins at natural step edges [9, 29]. It was also known from HRTEM that the graphene on the walls of miscut SiC steps appeared detached from the SiC step facets with a width set by the depth of the step (see figure 1.1) [9]. Since very shallow steps (a few nm deep) are easily achievable with current lithography, it was postulated and shown that narrow graphene ribbons can be grown from a pre-growth patterned SiC surface [23]. Since this early postulate, a good deal of work has been devoted to controlling where the ribbons grow, their edge type, details of their structure, and understanding the relationship between their structure and subsequent electronic structure and transport.

Controlling where EG ribbons grow and their edge type is the easiest step in sidewall ribbon production. This is done by lithographically pre-patterning trenches into the SiC(0001) surface with a predetermined edge direction and a trench depth [23, 27, 30, 31]. Once these trenches are heated in the correct Si vapor pressure environment, at the correct temperature, and for the correct time, single-layer graphene can grow on their refaceted sidewalls. Because of EG's epitaxy, the edge type of the EG ribbon that grows on the sidewall is determined by the original etched trench edge direction. When the trench is oriented perpendicular to the $\langle 1\bar{1}00 \rangle$ direction, the graphene that grows will have its AC-edge parallel with the step edge (see figures 1.1(e) and (g)). For convenience, we will refer to these SiC step edges as AC-edge steps or AC-facet walls. Conversely, when a SiC trench is oriented parallel to the $\langle 1\bar{1}00 \rangle$ direction, the graphene that grows has its ZZ-edge parallel with the step edge (see figures 1.1(g) and (h)). We will similarly refer to these SiC steps as ZZ-steps or facets.

Pregrowth patterning

Starting from a CMP polished SiC(0001) surface, a trench is etched by first coating the surface with a photoresist. ZEP520a is a positive resist for e-beam lithography with a fast exposure time that is used for shallower trenches. If deeper trenches are required, PMMA is used despite its exposure time being longer (and thus cost being higher). After the resist is developed, the exposed SiC areas are plasma etched using SF_6 or CF_4-O_2-Ar reactive ion etching to produce well defined trenches with step walls. The depth of the step is controlled by the etching current and time. Any remnants of the resist are removed prior to growing graphene. A poor etch or imperfect removal of the etch material after lithography will lead to poor EG sidewall growth (see section 1.2.3). The process details are found in [31]. Etching serves another important function. Unlike growth on natural steps, graphene does not grow well on the SiC(0001) plasma etched trench bottoms [32]. The etched trench bottom limits graphene growth primarily to the trench walls and small ribbons on the (0001) surface at the step edge. This statement is true as long as we are in a growth regime where the sidewall graphene thickness is less than two graphene layers.

Planarization

Natural steps on the SiC(0001) surface can be problematic in ribbon transport and device fabrication. These steps act as nucleation lines for graphene ribbons on the trench tops and facets. To prevent these spurious ribbons from growing on the

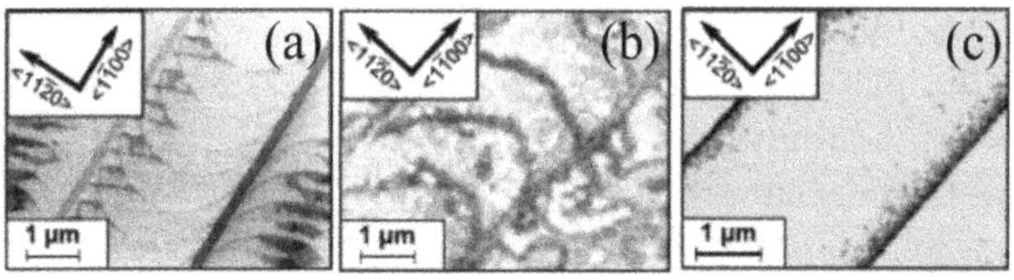

Figure 1.2. (a)–(c) SEM images of post graphene growth trenches. Graphene nucleation is seen as dark lines on the trench tops. The different samples were planarized before growth to give different terrace sizes Γ (a) $\Gamma = 500$ nm, (b) $\Gamma < 4\ \mu m$, and (c), $\Gamma > 20\ \mu m$. Reprinted from [36], with the permission of AIP Publishing.

(0001) surface, the SiC substrate can be planarized to produce step free (0001) surfaces. There are two methods that have been investigated: 'step flow control' and 'face-to-face' (FTF) growth. Step flow control uses pinning sites to trap steps at predetermined places. This was first demonstrated using pre-patterned mesas on SiC to produce 200 μm^2 step free surfaces [33]. A variation of this method has been reported that uses deposited amorphous carbon ring 'corals' to act as step traps [34]. The FTF method uses a second wafer of SiC placed above the sample SiC so their (0001) planes face each other [35]. This keeps the Si vapor pressure very high to flatten the (0001) surface without graphene growth.

Baringhaus *et al* [36] used the FTF method prior to patterning to essentially eliminate defect steps on the (0001) terrace tops after patterning. In their work, Baringhaus *et al* [36] found that planarization at 1300 °C for 5 h produces large terraces up to >20 μm. Heating to 1200 °C produces much smaller terraces. The reduction in the number of natural steps on the (0001) terrace tops is demonstrated in figure 1.2.

Graphene growth methods

Early EG growth on (0001) surfaces was done in an ultra high vacuum (UHV) [26]. However, the substrate disorder and film thickness variations after UHV growth were too high for both applications and basic research. Modern EG growth is done in a high Si-vapor environment to push growth nearer to equilibrium conditions, making graphene growth slower and more uniform. The slower growth makes thickness control easier. Emtsev *et al* [37] showed that growth in high Ar pressure kept the Si vapor pressure near the surface high so that the growth temperature raises well above the growth temperature in UHV. A second method, known as a confinement controlled silicon sublimation (CSS), places the SiC in a RF-heated carbon crucible [28]. The CSS method similarly keeps the Si vapor pressure high and leads to uniform thickness films. In both the Ar and the CSS methods, graphene growth is a function of temperature, time, and either Ar pressure and flow rate (in the Ar method), or crucible geometry in the CSS method. This means that growth temperatures given in this review are specific to the groups that produce them. The reader is therefore referred to the references provided for specifics. Details of the CSS crucible design and growth conditions can be found in [31, 38]. Finally, the FTF technique used to planarize samples has also been used to grow ribbons [36]. The face-to-face method is a simple version of the CSS method that can also increase the

Si-vapor pressure to slow growth. In Baringhaus *et al* [36], they simply increased the temperature in the planarizing process to 1500 °C over a minute and soaked for 10 min to grow ZZ-sidewall graphene.

The first carbon layer that grows on SiC(0001) is known as buffer graphene or the BG_0 layer [39]. The buffer layer is semiconducting [40]. When a second graphene layer begins to grow below the existing BG_0 layer, the buffer-substrate bonds are broken and replaced by van der Waals bonds to the newly formed buffer layer at the SiC surface. The bonding change makes the original buffer layer metallic (thicker films can be grown but will not be discussed here). We mention this because graphene's top down growth on SiC affects how the sidewall graphene terminates into the macroscopic (0001) surface. Depending on growth conditions, the sidewall graphene can terminate into either metallic or semiconducting graphene on the (0001) surface. This will be discussed in more detail in section 1.3. The two possible ribbon edge terminations demonstrate that sidewall ribbons are not the idealized termination used in most calculations for the electronic structure of graphene ribbons.

Besides the obvious growth parameters mentioned above, the SiC substrate polytype plays an important role in sidewall ribbon growth. While polytype may not seem important, it plays a pivotal role in ZZ-ribbon growth as will be discussed in section 1.2.3. For ZZ-ribbons, the polytype determines how the graphene bonds to the facet walls and ultimately the ribbon width distribution on the faceted sidewalls.

1.2.2 Graphene ribbons on vicinal SiC

Before discussing ribbon growth on patterned SiC, we briefly discuss growth on miscut (or off-axis) SiC surfaces. Off-axis SiC(0001) contains parallel step arrays that are used to nucleate parallel graphene ribbons. SiC can be purchased with an off-axis cut that is usually 3.5°, 4° or 8° and can be oriented to give both AC- and ZZ-step arrays. A number of works have used the parallel natural steps on miscut SiC(0001) to grow graphene and graphene ribbons on both AC and ZZ steps [24, 41–44]. Growth has been done on 4H ZZ-arrays in a 900 mbar Ar furnace at 1600 °C–1700 °C [43, 44] and on 6H AC-arrays in a Si seeded N_2 atmosphere [41]. In these studies, the graphene film consist of multiple layers.

A more interesting growth study was performed by Kajiwara *et al* [24]. They studied growth on 4° off axis AC-steps on 6H SiC. Figures 1.3(a) and (b) show the step structure. Rather than growing by Si sublimation, they grew graphene using a thermal C-deposition source (operating at 2000 °C). This causes the semiconducting buffer graphene to grow on the nano (0001) surfaces instead of on the $\{1\bar{1}0n\}$ facets. This is in contrast to all the sublimation methods where graphene growth begins on the facet planes. In the facet structure in figure 1.3(a), the width of each (0001) terrace is ~10 nm. After growth, the sample is H_2-passivated according to the process of Sforzini *et al* [45]. The passivation breaks the graphene-Si bonds and converts the semiconducting graphene to metallic monolayer (ML) graphene. In the GW approximation, for AC-ribbons with symmetric edge termination, a 10 nm width would predict a band gap of 0.14–0.43 eV depending on the ribbon type [6].

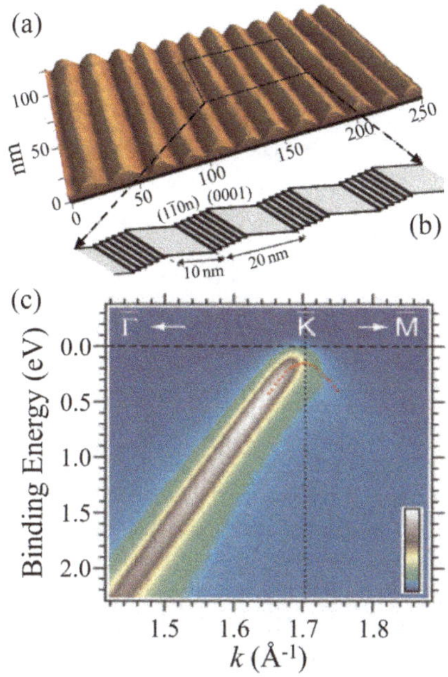

Figure 1.3. (a) AFM image of the 4° vicinal SiC surface after hydrogen etching, showing the periodic terraces and facet structure. (b) A schematic drawing of the SiC surface in (a). (c) ARPES intensity map along the graphene Γ-K-M line in the (0001) plane. The intensity map shows the linear π-band dispersion of single-layer graphene. Red dashed lines show the folding of the valence band 0.14 eV below E_F. Reprinted with permission from [24], Copyright (2013) by the American Physical Society.

Indeed, ARPES measurements, shown in figure 1.3(c), find a gap of 0.14 eV. We believe this technique will have important implications for narrow graphene ribbon growth.

1.2.3 Sidewall graphene growth

Growing graphene on the trench walls is a high temperature process involving significant mass flow that will change the wall geometry. Controlling the facet shape was an early challenge requiring a balancing act between controlling the graphene thickness, limiting graphene overgrowth on the trench tops, and preventing the trenches from melting into the low energy (0001) surface. An important parameter in sidewall graphene growth, only recently discovered, is the SiC polytype. It was assumed early on that ribbon growth on steps of 4H- and 6H-SiC would be very similar. It turns out, however, that their energetics are very different. Figure 1.4 shows the surface free energy as a function of angle, $F^{\mathrm{SiC}}(\theta)$, for both 4H- and 6H-SiC. The plots include facet angles for both AC $\{1\bar{1}0n\}$ and ZZ $\{11\bar{2}n\}$ planes. Two different temperatures are plotted. Minima in $F^{\mathrm{SiC}}(\theta)$ can correspond to stable facets. Entropy causes the minima to become small and the facet unstable in the range of temperatures where graphene grows (>1450 °C). Graphene ribbon growth therefore becomes problematic because the trench structure can melt during growth, as will be shown in the next section.

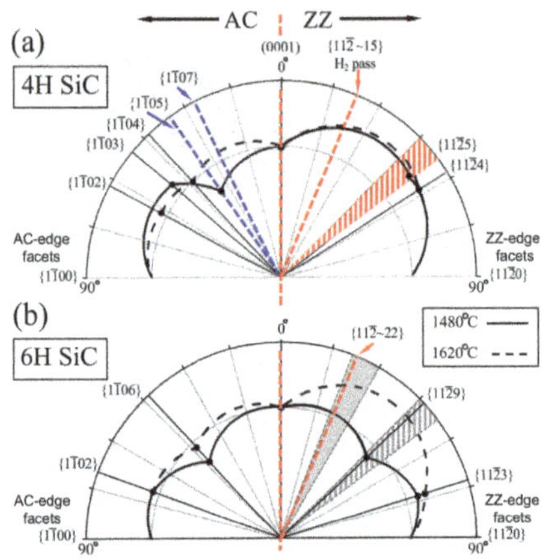

Figure 1.4. Surface free energy, $F^{SiC}(\theta)$, plots for (a) 4H- and (b) 6H-SiC based on calculated (black lines) and experimental (dots) growth rates for two temperatures. Reprinted from [46], Copyright (1999), with permission from Elsevier. Both 4H- and 6H-SiC polytypes have pronounced minima for $\{1\bar{1}0n\}$ facets, indicating stable AC-facets. In contrast, there is no $F^{SiC}(\theta)$ minimum for 4H-ZZ-facets. Only 6H-SiC has stable ZZ-facets, where $F(\theta)$ has a minimum for the $\{11\bar{2}9\}_{6H}$ facet. Blue dashed lines mark the 4H-AC sidewall graphene facet angles measured by ARPES in previous works from [47, 48]. Red dashed lines mark the stable ZZ-facets from H_2 passivated 4H- and 6H-ribbons, respectively. Gray shaded area in (b) is the estimated facet angles from STM measurements of graphene growing on 6H SiC steps in [50]. The grey hatched areas in (b) show a range of angles where disordered ZZ-sidewall graphene grows [50].

Figure 1.4 shows well defined minima for AC $\{1\bar{1}0n\}$ facets for both 6H and 4H polytypes. Therefore, there are stable AC-facets on both polytypes. In contrast, $F^{SiC}(\theta)$ for ZZ-facets predicts that only the 6H-SiC polytype will have a stable facet; the $\{11\bar{2}9\}_{6H}$ at $\theta_F = 47.5°$. It is therefore clear that, at least for ZZ-facet walls, the polytype is important for facet stability. Of course, we do not expect $F^{SiC}(\theta)$ for bare SiC to be the same as $F^{G+SiC}(\theta)$ once graphene has grown on the sidewall. Figure 1.4 simply shows that it is not unreasonable that there can be differences in ZZ-sidewall graphene grown on the two polytypes. As we will show in section 1.2.3, not only is the facet structure on AC- and ZZ-ribbons very different, the polytype plays a critical role in the structure and the graphene–SiC interaction for ZZ-edge ribbons. We will therefore discuss the details of AC- and ZZ-edge ribbons growth separately.

AC sidewall graphene growth

While atomic force microscopy (AFM), electrostatic force microscopy (EFM), or lateral force microscopy (LFM) can give quick information on EG-sidewall growth, there are serious limits on their ability to make quantitative measurements. These techniques miss many of the details of the sidewall structure and electronic properties of this complicated system. For example, scanning transmission electron microscopy (STEM) and angular resolved photoemission electron microscopy (ARPES) are sensitive to the number of graphene layers. Both techniques have shown that, for the same growth recipe, AC sidewall graphene will either be a single layer or a disordered mixture of single and double layers if the initial trench etch

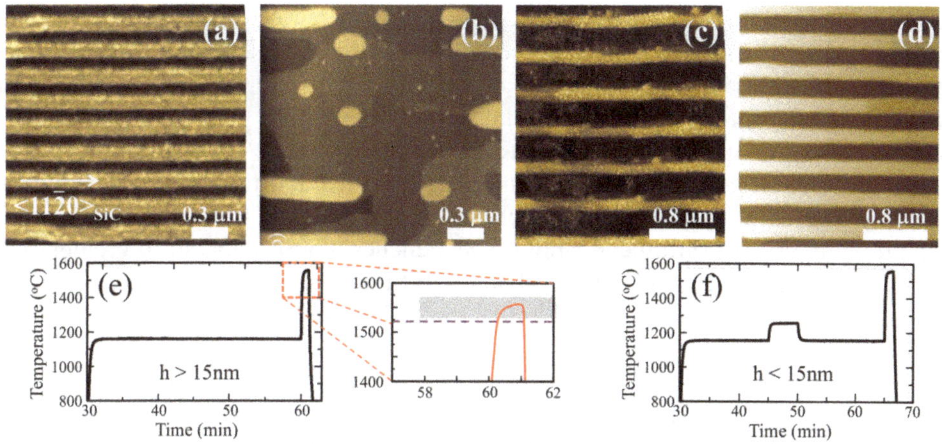

Figure 1.5. (a) AFM image of 25 nm deep AC trenches etched in 4H SiC(0001). (b) AFM image of the trenches in (a) after heating directly to the graphitization temperature (1520 °C for 1 min) in a CSS furnace. (c) Same as (b) except trenches were annealed at 1150 °C for 30 min before graphitization. (d) Same as (c) except graphitization temperature was between 1530 °C–1565 °C. (e) AC-ribbon growth temperature cycle for trenches deeper than 15 nm (inset shows details near T_G). Dashed line is the graphene growth temperature for ribbons in (c). Gray region is the growth temperature for ribbons in (d). (f) Growth cycle for sub 15 nm trench ribbons.

lithography is done improperly [31]. This level of information is simply not available in scanning probe microscopies. Indeed, the recipes for optimum ribbon growth given in this section and in section 1.2.3 are a result of an extensive body of comparative work using ARPES, scanning tunneling spectroscopy (STS), STEM, low energy electron and x-ray photoelectron emission microscopy (LEEM/XPEEM) experiments. In this section, we discuss the state of the art understanding of AC sidewall ribbons based on all of these studies.

It was discovered early on that annealing AC trenches at 1150 °C before reaching the graphene growth temperature (T_G) were necessary to both help prevent the trench structure from melting and to reduce fluctuations of the trench step edges [51]. This is demonstrated in figure 1.5 where we compare post growth step structures with and without an annealing step prior to graphene growth in a CSS furnace. Figure 1.5(a) shows an AFM image of a 24 nm deep trench array before growth. Without a pre-annealing step, the trench structure has completely melted, as shown in figure 1.5(b). Annealing at T_A = 1150 °C for 30 min before EG growth preserves the trench structure, as shown in figures 1.5(e) and (d). It is believed that the low temperature anneal allows a free energy minimum in figure 1.4(a) to develop so that a stable facet can form.

However, at T_G, between 1480 °C–1620 °C, even a stable facet will melt as figures 1.4(a) and (b) show. It was found, however, that rapid heating from the anneal temperature to EG's growth temperature prevented facet melting. It is believed that rapid growth of graphene on the facet walls prevents further evaporation of Si, thus stabilizing the facet at high temperatures. The reasoning for this statement is much the same as why growing more than one graphene layer on the (0001) surface is inhibited unless the growth temperature is increased (i.e. graphene growth is an activated process where the energy barrier is a function of graphene thickness). A similar method, known as carbon capping, is used to stabilize ion implanted dopants from evaporating at the SiC surface [52].

There are subtleties in both T_G and the annealing cycle that affect the ribbon structure. Growing too hot or too long leads to the sidewall EG overgrowing the facet up and onto the SiC(0001) trench tops. The preferred CSS temperature cycle for growing ML AC-graphene ribbons on trenches whose depths are greater than 15 nm is shown in figure 1.5(e) [31]. The ramp time from the anneal temperature to the growth temperature (T_G = 1530 °C − 1565 °C) is ~30 − 40*s*. The growth time is ~50 − 70*s*. ARPES and STEM have determined that growing above ~1575 °C causes an additional bi-layer graphene layer to grow above the ML sidewall graphene. A slightly higher T_G ($\Delta T \sim 30$ °C) is the difference between the wandering step edges in figure 1.5(e) and the much straighter edges in figure 1.5(d) [31]. It has been empirically found that sub-15 nm trenches require an additional annealing step at higher temperature before graphene growth (see figure 1.5(f)). The variability in temperature and time given above are due to aging of the carbon crucible in the CSS furnace. The furnace walls develop a SiC layer after repeated use that changes the Si vapor pressure for the same oven temperature. Usually, the furnace is outgassed at 1650 °C for 15 min with a sacrificial SiC wafer inside the crucible. The conditioning is done after every five to ten samples. Test ribbons are checked periodically with AFM/EFM to test for overgrowth so that the furnace temperature can be slightly adjusted to maintain ribbon quality. More details about ribbon quality, layer thickness, and doping as a function of growth temperature will be discussed below when the ARPES results are discussed.

A single layer of graphene may flow over a ~1 nm high natural AC-step in a smooth way, but for taller steps the facet wall structure and thus the graphene ribbon structure are more complicated. Figure 1.6(a) shows a high-resolution STEM image of an AC-step after one graphene layer has grown on the faceted trench sidewall. The step facet structure consists of pairs of $(1\bar{1}05)$-(0001) nano-steps near the top and bottom of the trench and a large $(1\bar{1}07)$ facet in the center of the sidewall. The facet angles, θ_F, for the $(1\bar{1}05)$ and $(1\bar{1}07)$ planes are 37.1° and 28.4°, respectively. We note that the $(1\bar{1}07)$ facet area is larger than the $(1\bar{1}05)$ area. We would therefore expect that the minimum in $F(\theta)$ for the $(1\bar{1}07)$ facet should be deeper than the minimum for the $(1\bar{1}05)$ facet. This is opposite the prediction for $F(\theta)$ in figure 1.4(a). This observation demonstrates that $F^{SiC+G}(\theta) \neq F^{SiC}(\theta)$ in figure 1.4(a). The total AC-ribbon width is about twice the pre-growth trench depth because of the facet angle. For the case of a single sidewall graphene layer, the graphene terminates into (0001) terrace top as a buffer layer (see figure 1.6(b)). On the larger $(1\bar{1}07)$ facets, the STEM image in figure 1.6(a) along with the STM images in figures 1.6(d) and (e) show that the graphene-SiC spacing undulates over the facet. In other words, the graphene is periodically bonded to the SiC facet [48]. The average graphene-SiC separation for the rippled graphene on the $(1\bar{1}07)$ facet is 3.5 Å; slightly larger than the 3.336 Å interlayer spacing of graphite [53].

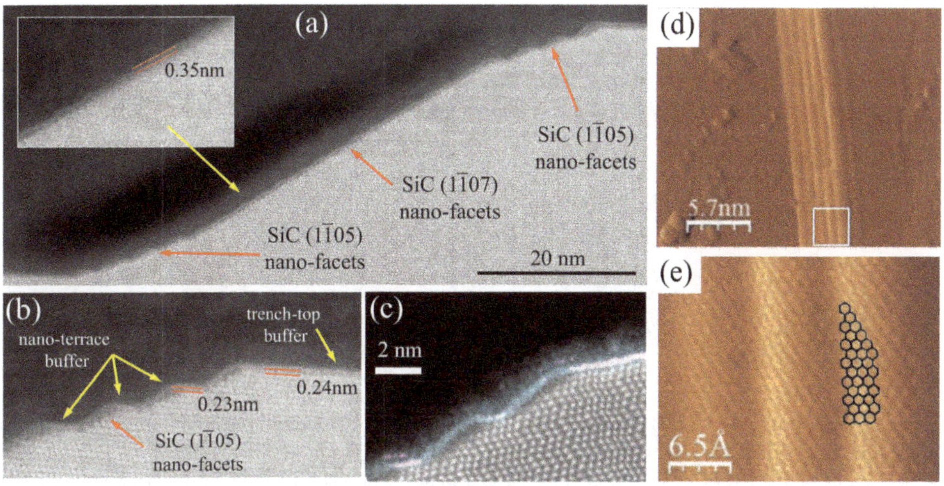

Figure 1.6. Overview of high resolution STEM of an AC-trench. (a) STEM after a single graphene layer has grown. The vertical trench has broken into (1$\bar{1}$05) nano-facets at the trench top and bottom and a large (1$\bar{1}$07) facet in the trench center. (b) Closeup of the (1$\bar{1}$05) nano-facets near the trench top. (c) Pseudocolor image of the sidewall graphene layer based on STEM-EELS analysis. Cyan highlights electronic graphene and white shows graphene perturbed by strong interactions with the SiC below. (d) STM constant current image of sidewall graphene showing 3.3 nm high undulations. (e) Blowup of the rectangle in (d) showing that the AC-edge of the ribbon runs parallel to the sidewall edge. Reproduced from [48]. © IOP Publishing Ltd and Deutsche Physikalische Gesellschaf. All rights reserved.

On the smaller (1$\bar{1}$05) – (0001) nano-step regions shown in figure 1.6(b), the graphene is detached from the SiC (1$\bar{1}$05) plane and then transitions onto the flat nano-(0001) terrace. The 0.23 nm graphene-SiC spacing on the nano-(0001) terraces is slightly shorter than the buffer graphene-SiC spacing on the large (0001) terrace tops (0.24 nm) (see figure 1.6(b)). The similar spacings imply that graphene on the (0001) nano-terrace should be electronically similar to the semiconducting buffer graphene layer on the macroscopic (0001) surface. Palacio *et al* showed that the graphene on the nano-(0001) terrace is electronically very different from the graphene on the (1$\bar{1}$05) facet but similar to the semiconducting buffer graphene. This is demonstrated in figure 1.6(c), which uses an electron loss component associated with graphene to make a false color image in STEM-EELS. The graphene on both the nano-terraces and the macroscopic (0001) flat surface is bright, indicating that they are electronically similar. In other words, the (1$\bar{1}$05) metallic AC nano ribbons are bounded on either side by semiconducting buffer graphene on the nano (0001) planes. This structure has been predicted to open a ~1 eV quantum confinement gap [48]. This prediction remains to be tested.

ARPES has been used to measure the band structure of AC EG-sidewall ribbons. While a single ribbon cannot be measured using typical ARPES beam diameters, ~40 μm, a patterned array with a large number of ribbons can be measured as long as the ribbons are straight, ordered and coherent. This is the case with patterned sidewall ribbons where the average ARPES spectra represent a very good

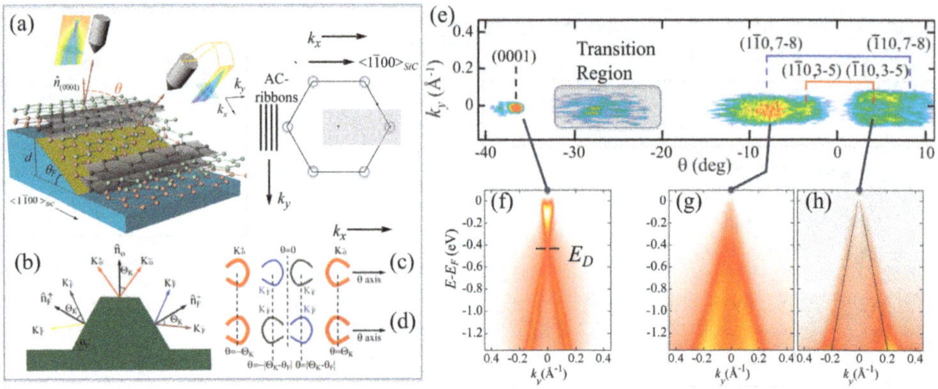

Figure 1.7. Schematic of a faceted side wall (tilted by an angle θ_F) with two graphene layers. Grey regions are curved graphene. The detector is rotated by θ relative to the (0001) surface normal to accept electrons with momentum along the facet planes's $\Gamma - K$ direction (scan along the shaded region of the facet graphene's BZ in (a)). Opposing facets give rise to a pair of plus and minus K-points (see (b)). (c), (d) The Dirac cones from the facets appear along θ at angles given by $\theta = \pm|\theta_F \pm \theta_K|$. (e) ARPES constant E map from the trench array after graphitization (AC edges parallel to k_y and $E - E_F = -0.43$ eV). (f)–(h) k_y-cuts through the Dirac cones of the (0001) and the facet surfaces. (f) Dirac cone of the (0001) surface, (g) the $(1\bar{1}07)$ facet, and (h) the $(1\bar{1}05)$ facet. (e)–(h) Reprinted by permission from Nature Physics [47], © 2013 Springer Nature Publishing AG.

representation of a single sidewall graphene ribbon. Because ARPES measures the band dispersion $E(k_{\|})$, where $k_{\|}$ is in the plane of the local surface normal, it is possible to separate the band structure of graphene on the (0001) surface from the graphene on the facets. Figure 1.7(a) shows a schematic of how ARPES measures the band structure from graphene on the sidewall facet. For AC-facets, we define k_x and k_y as the in-plane momentum of the facet surface parallel and perpendicular to (Γ-K), respectively, i.e. k_y is parallel to the AC-step (see schematic BZ in figure 1.7(a)). In principle, the facet angle can be determined from the θ positions of the facet cones. However, this must be done with care because the Fermi surface has an ambiguity caused by the existence of *two* opposing facets from the trench array [31]. The problem is shown schematically in figures 1.7(b)–(d). The Dirac cones at the K-points are tilted from the local surface normal by $\pm\Theta_K$. Opposing facets can each produce a Dirac cone along the θ-axis between the (0001) K_o^+ and K_o^- points. The angular positions of the facet cones are then $\theta = \pm|\Theta_K - \theta_f|$. The θ-positions of the opposing facet cones therefore depend on the magnitude of θ_F, as shown in figures 1.7(c) and (d). The ambiguity in determining θ_F can be removed by noting the symmetry of the Dirac cone. Since the Dirac cone is distorted from a perfect circle by diffraction effects [54, 55], the correct tilt angle of the cone can be determined by comparing its distortion with the distortion of the Dirac cone from the (0001) surface (see figures 1.7(c) and (d)).

Figure 1.7(b) shows the first ARPES partial-Fermi surface from an AC-ribbon array (30 nm trench depth) grown from the earliest growth recipe. Even these earliest growth protocols showed reasonably ordered Dirac cones corresponding to $(1\bar{1}07)$ and $(1\bar{1}03)$ facets [47]. Note that the facet plane index n, $(1\bar{1}0n)$, show a range of possible values in figure 1.7(b). This represents the measurement angular uncertainty. We note that the misidentification of the $(1\bar{1}05)$ facet as $(1\bar{1}03)$ in the early ARPES measurements was due in part to the angular uncertainty caused by the the sample order. Figures 1.7(c)–(e) show cuts through the Dirac cones from the (0001), $(1\bar{1}07)$, and $(1\bar{1}05)$ surfaces, respectively. The facet cones are reasonably resolved with some broadening in the k_y direction. The surface normals of the $(1\bar{1}07)$ and $(1\bar{1}05)$ AC-facets are tilted $\theta_7 = 28.4°$ and $\theta_5 = 37.1°$ relative to the (0001) surface, respectively. The Dirac point, E_D, of the facet cones is shifted by about +0.43 eV relative to the (0001) surface Dirac point. The shift indicates that the facet cones are p-doped relative to the (0001) graphene. As we will now discuss, the shift is related to the difference in the number of graphene layers on the (0001) compared to the $(1\bar{1}0n)$ facets and not due to an intrinsic doping change.

More detailed studies of 4H facet cones, subsequent to these early ARPES measurements, have given more information about their growth, electronic structure and long range order. The early sidewall ribbons used in figure 1.7(b) were grown using a lower growth temperature compared to the current methods described in figure 1.5. While AFM and EFM images of AC-ribbons grown at slightly different temperatures look similar, there are substantive temperature effects in sidewall graphene growth. First, the facet Dirac cones appear more disordered compared to the Dirac cone of the flat (0001) surface. A comparison of the π-bands and their momentum distribution curves (MDCs) is shown in figure 1.8 for two different growth temperatures. Figures 1.8(a) and (b) show the bands from the (0001) surface. The (0001) cones are well defined with a k_y broadening, $\Delta k_y = 0.07$ Å^{-1}. When the ribbons are grown at the low temperature of $T_G = 1520$ °C, the Dirac cones of the $(1\bar{1}05)$ and $(1\bar{1}07)$ facets are approximately twice as broad, $\Delta k_y = 0.14$ Å^{-1}, as the (0001) Dirac cones (see figures 1.8(c) and (d)). Some of the broadening is from macroscopic variations of the step edges as seen in figure 1.5(c) [56], and some comes from the microscopic rippling of the $(1\bar{1}07)$ graphene (see STEM images in figure 1.6(a)). The latter leads to a variation in surface normals and thus a variation in k_y. The broadening corresponds to an approximate long range order of $L = 2\pi/\Delta k \approx 5$ nm. Using the higher temperature growth recipe ($T_G = 1550$ °C) shown in figure 1.5(e), leads to more ordered facet Dirac cones. This recipe leads to much smaller Δk_y that are nearly identical to those of the (0001) surface graphene.

Note that the doping, $E_D - E_F$, of both $(1\bar{1}05)$ and $(1\bar{1}07)$ facets depends on T_G. At the lower T_G, the facet graphene doping is essentially zero. Ordered ML graphene on the (0001) surface is normally n-doped by −0.45 eV (see figure 1.8(a)). Doping is reduced when the graphene has some amount of adsorbates [45, 57]. Growing at the higher T_G not only orders the ribbons, it also increases their doping (∼−0.24 eV) closer to the value of clean graphene.

Finally, the optimized growth temperatures in figures 1.5(e) and (f) give only ML graphene. A typical facet Dirac cone spectrum is shown in figure 1.9(a). Growing 10 °C hotter (1575 °C) leads to bi-layer growth. At these higher growth temperatures, the facet Dirac cone spectrum shows a mixture of predominantly bi-layer graphene with some ML (see figure 1.9(b)). At the same time bi-layer grows on the facet, ML grows

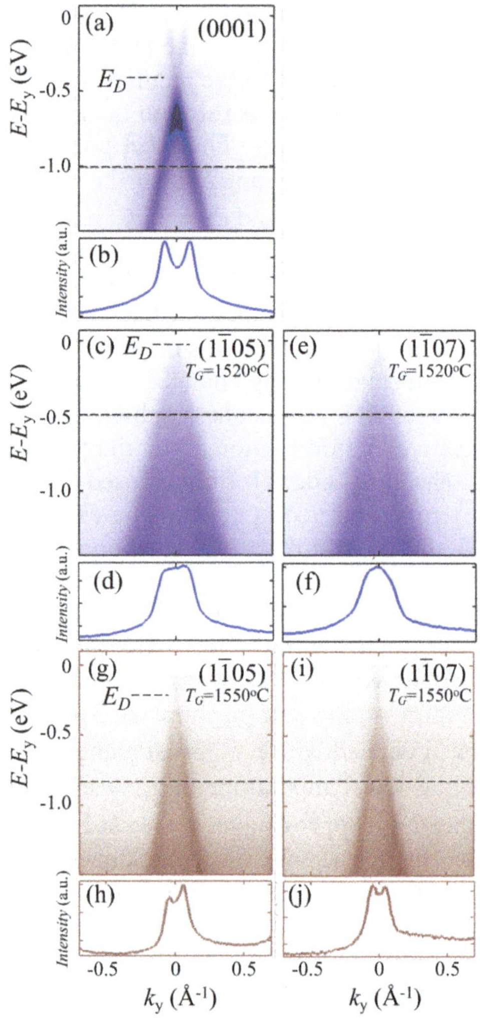

Figure 1.8. A comparison of the facet cone order for the different growth recipes in figure 1.5(e) and (d). For all data, $h\nu = 36$ eV and the MDCs were taken at $E - E_D = -0.5$ eV (dashed lines). Upper panels are ARPES cuts through the K-point ($k_y = 0.0$ Å^{-1}) and lower panels are MDCs through the dashed line in the upper panels. (a), (b) Results for (0001) ML graphene, $E_D = -0.43$ eV. The Dirac cone broadening is $\Delta k_y = 0.07$ Å^{-1}. (c)–(f) Similar data sets as (a) and (b) after 1520 °C graphene growth. The Δk_y width of the (1$\bar{1}$05) and (1$\bar{1}$07) cones are 0.12 Å^{-1} and 0.14 Å^{-1}, respectively. (g)–(j) The same cuts as (c)–(f) but the sample was grown at $T_G = 1550$ °C. The Δk_y widths of the (1$\bar{1}$05) and (1$\bar{1}$07) cones are now 0.08 Å^{-1} and 0.06 Å^{-1}, respectively.

on the top of the trenches. The (0001) ML can short the semiconducting buffer in transport measurements.

ZZ sidewall graphene growth

The growth of ZZ-edge graphene ribbons has proven to be more complicated than AC-ribbons. For more than four years, it remained a mystery as to why some groups

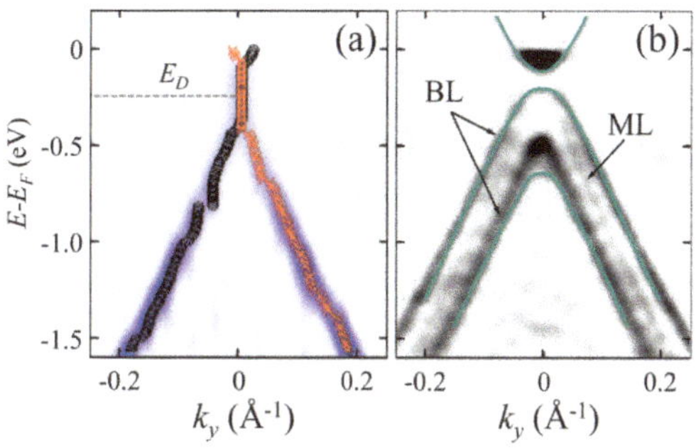

Figure 1.9. (a) Negative second derivative image of a ML AC-sidewall graphene cone using the recipe in figure 1.5(e) with T_G = 1565 °C. Red/black circles mark the maxima MDC intensities. (b) AC sidewall Dirac cone grown using the same recipe in (a) except that T_G is raised 10 °C to 1575 °C. The green overlayers are theoretical bands for a pure bi-layer film. A weak ML band is also marked.

were able to grow ZZ-edge sidewall graphene on SiC [50], find evidence of their predicted edge states [50], and measure their ballistic transport [58, 59], while other groups found no evidence that metallic graphene even grew on the ZZ-edge sidewall. It is now known that metallic ZZ-edge EG only grows on the facet walls of the 6H SiC polytype, not the 4H [49]. ZZ-edge ribbons do grow on the 4H polytype but their bonding to the SiC facets severely distorts the facet graphene's π-bands so that the ribbons are not metallic. The experimental observed difference between 4H and 6H $\{11\bar{2}n\}$ ZZ-facet growth are in fact expected. As the free energy plots in figures 1.4(a) and (b) show, 6H SiC has a predicted stable ZZ-facet, while 4H SiC does not. In this section, we show what is now known about the structural differences of ZZ-ribbons for the two polytypes. Because growth on the two polytypes is different, we will discuss 4H- and 6H-ZZ-graphene ribbons growth separately.

4H-ZZ-edge ribbons: There have been several attempts to grow 4H-ZZ EG. The first used the AC growth ribbons recipe in figure 1.5(e). Growth times were extended to the point where the step pattern melted without evidence of sidewall graphene Dirac cones. In a second attempt, a buffer layer film was grown on the sample before step patterning. The idea for this method was to cap the tops of the steps to prevent Si evaporation and thus allow higher growth temperatures without pattern melting and at the same time offer a seed graphene layer that would grow over the sides to make ZZ-sidewall graphene. In both cases, there was no clear indication that stable facets form, let alone that graphene had grown on them. This was demonstrated by LEEM, XPEEM, and ARPES.

LEEM measurements found no diffraction rods that should be associated with ZZ-oriented $\{11\bar{2}n\}$ facets. Instead, only diffraction rods from AC $\{1\bar{1}0n\}$ facets were found. This is shown by comparing LEEM images of AC- and ZZ-ribbons (see figures 1.10(a) and (b)). The LEEM results suggest that the ZZ-edge facets are disordered and have partially broken into nano-AC-facets (see figure 1.10(c)). A similar instability has been observed in miscut 6H SiC [60, 61] and in the faceting of large circular SiC islands [62]. In fact, it was suggested that the apparent lack of 4H

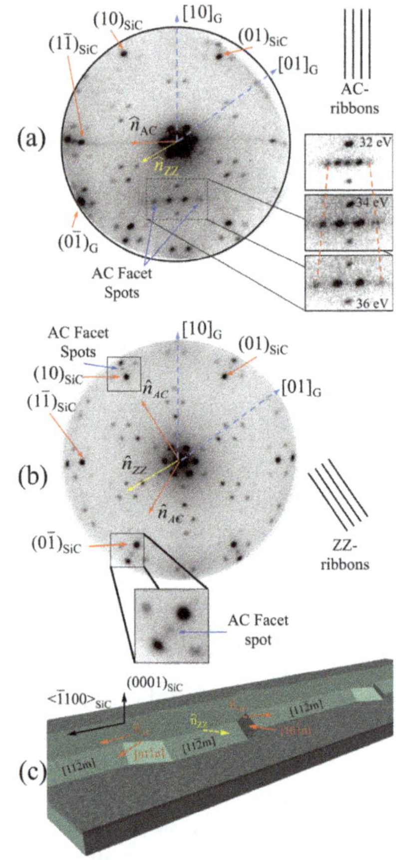

Figure 1.10. (a) μ-LEED image of AC-ribbons (E = 32 eV). The normals to ZZ and AC-step edges are marked. The SiC $(0\bar{1})$ rods from an AC $(10\bar{1}n)$ facet are clearly visible in the blowup of the $(0\bar{1})_{SiC}$ rod region. The blowup shows that the facet rods move in the $\hat{n}_{AC}$ direction as expected for AC-facets. (b) μ-LEED image of ZZ-ribbons (E = 32 eV). (c) Schematic of the ZZ-step faceting into AC nano-facts.

ZZ-ribbon Dirac cones in ARPES measurements was due to this edge disorder that would broaden the cones in Δk and thus suppress their appearance [63]. This assertion, however, is wrong for two reasons. First, it ignores XPEEM studies of 4H ZZ-ribbons that are insensitive to edge order. The XPEEM showed that non-metallic graphene was present on the facets. Secondly, ARPES measurements of H_2 intercalated 4H ZZ-ribbons have also shown that graphene grows on the sides of 4H-sidewalls, but unlike 6H ZZ-sidewalls, ZZ-ribbons on 4H SiC are strongly bonded to the SiC facet walls.

XPEEM measurements first showed that graphitic carbon exists on the 4H-ZZ sidewalls [32]. Functionalized graphene, i.e. sp^3 bonded carbon, has a unique C 1 s spectrum compared to metallic graphene. Figures 1.11(a) and (b) show typical spectra from functionalized buffer graphene and metallic ML graphene layer on SiC (0001). When ML is present, the C 1 s spectrum has a large S_{ML} component near a binding energy (BE) of 284.5 eV. When only functional buffer graphene is present,

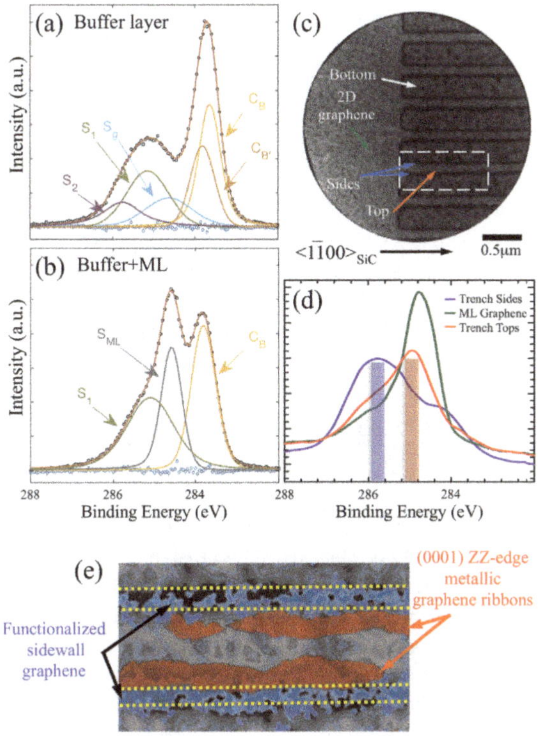

Figure 1.11. C 1 s spectra from buffer graphene (a) and ML graphene above the buffer (b) ($h\nu$ = 2,514 eV). The buffer spectra is fit with five components; two for functionalized buffer carbon (S_1 and S_2), one for sp^2 buffer carbon (SG), and two peaks for SiC carbon (C_B and C'_B). The ML spectra are described by three components; a buffer component (S_1), the ML graphene (S_{ML}), and a single SiC carbon peak C_B [38]. (c) A DF-LEEM image of the ZZ-step array. (d) XPEEM C 1 s spectra taken at on different parts of the ZZ-step array in (c) ($h\nu$ = 500 eV). Blue is on trench sides indicating that the sidewall buffer graphene. Red is metallic graphene on trench tops near the step edge. (e) An overlay image of the dashed rectangle region in (c). Grey is the DF-LEEM image of the steps. Blue (red) are XPEEM images using the C 1 s energies of buffer (blue region in (d)) and ML (red region in (d)) for contrast. Yellow dotted line shows the facet wall boundaries.

the C 1 s spectral weight is shifted to higher BE due to the larger sp^3 bonded S_1 peak at BE = 285.2 eV [39, 64]. Figure 1.10(d) shows XPEEM C 1 s spectra from three places on the step array in figure 1.10(c): the 2D ML graphene region on the unpatterned part of the sample, on the (0001) step tops near the step edges, and on the trench sides. It is clear that the trench sides have C 1 s spectra more typical of functionalized buffer graphene and that the trench tops are more ML in character. To demonstrate the location of both functionalized and ML graphene, two XPEEM contrast images were made on the same region as the dark field (DF) LEEM image in figure 1.11(c); one using a photo-electron pass energy of 285 eV for the ML and the other using BE = 285.8 eV for functionalized buffer graphene (the red and blue energy window in figure 1.11(d)). The XPEEM images overlaid on the LEEM trench image are shown in figure 1.11(e). The dotted lines parallel to the top edge are the boundaries of the facet walls. The composite image clearly shows metallic

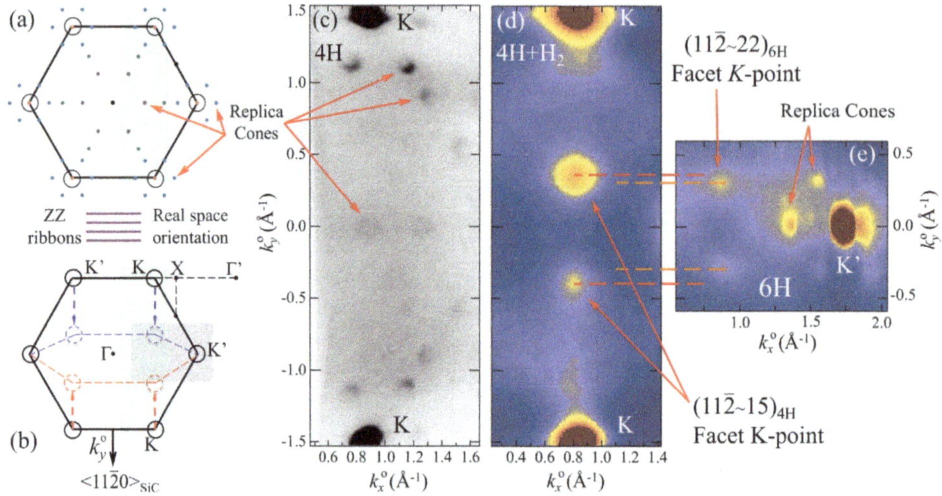

Figure 1.12. (a), (b) Graphene (0001) BZ. Dots in (a) are 6 × 6 replica cones. (b) The compressed graphene BZs (dashed lines and circles) of the $\{11\bar{2}n\}$ (blue) and $\{\bar{1}\bar{1}2n\}$ (red) plotted in the (0001) coordinate frame. (c)–(e) ARPES constant E cuts for different ZZ-sidewall samples; $E - E_F = 0.09$ eV and $h\nu = 36$ eV. (c) 4H ZZ-ribbons scan (green area in (a) and (b)) showing (0001) cones and their (6 × 6) replicas. (d) Same as (c) but after H_2-passivation. Dirac cones appear from $\{11\bar{2} \sim 15\}_{4H}$ facets. (e) 6H ZZ-ribbons scan (blue area in (b)) (no passivation). The cones are from graphene on $\{11\bar{2} \sim 22\}_{6H}$ facets. Horizontal lines show that the cones from 4H and 6H polytypes occur at different θ_F. Reprinted with permission from [49], Copyright (2019) by the American Physical Society.

ZZ-graphene ribbons grow on the (0001) trench top near the step edges. No appreciable amount of metallic graphene grows over and onto the facet walls. Instead, the facets are covered with a functionalized graphene layer. In other words, graphene does grow on the 4H facet walls but it is so strongly bonded to the SiC that its π-bands are severely distorted to the point that they do not show up in ARPES.

ARPES measurements have confirmed that metallic graphene does not grow on the 4H-ZZ sidewalls. A typical ARPES Fermi surface (FS) from 4H ZZ-ribbons grown in a CSS furnace is shown in figure 1.12(c). The FS only shows the Dirac cones from the (0001) surface along with a series of 6th-order replica cones associated with the reconstructed graphene-SiC (0001) interface (see figure 1.12(a)). No Dirac cones from metallic graphene from the facets are visible along the line from K and K as would be expected (see figure 1.12(b)). However, when the same sample is passivated at 900 °C for 1 h in a 800 mTorr H_2-atmosphere, the ARPES FS shows clear Dirac cones tilted relative to the K-point Dirac cones of the (0001) surface (see figure 1.12(d)). The cones occur at a facet angle of $\sim 23 \pm 1°$ which would correspond to $\{11\bar{2} \sim 15\}_{4H}$ ($\theta_F = 23.6°$) planes. H_2 passivation is known to break the graphene-Si bonds that make the buffer graphene layer on SiC (0001) semiconducting. In other words, ZZ-edge graphene does grow on $\{11\bar{2}n\}$ planes but it is so strongly bonded to the facet walls that it is essentially an insulator.

The ARPES measured band structure of the H_2 passivated ZZ-ribbons is shown in figure 1.13. The band are shown along the K-K'-Γ′ direction. The positions of the

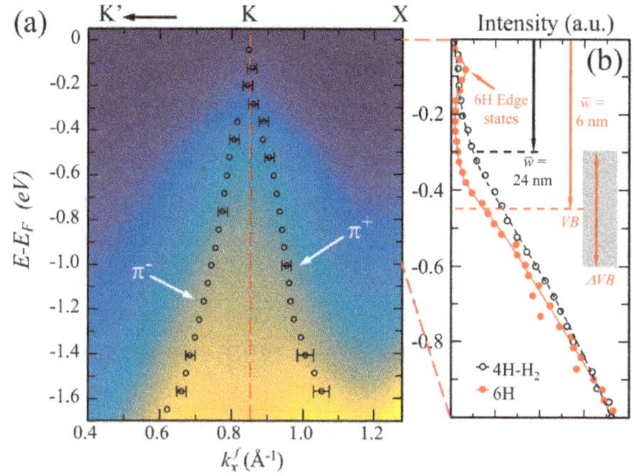

Figure 1.13. (a) ARPES bands from H^2-passivated 4H-ZZ-ribbons along the K'K Γ′ direction. k_x^f is in the $(11\bar{2} \sim 15)_{4H}$ plane. (b) Integrated π-band intensity from MDCs for H_2-passivated 4H ribbons (black circles) and 6H ZZ-ribbons (red circles). Red arrows mark the estimated valance band edge and width from STM data for 6H ribbons. Reprinted with permission from [49], Copyright (2019) by the American Physical Society.

4H π-bands in figure 1.10(a) are marked by circles. The H_2-passivated 4H ribbon's π^- and π^+-bands are nearly symmetric about the K-point with the same band velocity found in macroscopic sheets, $v_F \sim 1 \times 10^6$ m s^{-1}. Miettinen *et al* [49] show that the H_2-passivated 4H ribbons are wide. The ribbons are essentially the width of the facet wall, $\overline{W}_{4H} > 24$ nm, and they show no evidence of the predicted ZZ-ribbon edges states or the energy gap predicted for small ZZ-ribbons [6].

6H-ZZ-edge ribbons: While metallic ZZ-edge graphene does not grow on the facet walls of 4H SiC, it does grow on 6H SiC facets. Work on 6H-ZZ-ribbons has been reported by two groups. The first group to study 6H-ZZ sidewall ribbon growth was Baringhaus *et al* [50]. They grew their ZZ-ribbons starting from planarized 6H samples as described in section 1.2.1. After patterning, the sidewall graphene was grown using the face-to-face (FTF) method also described in section 1.2.1. In a later paper [58], the growth was different; samples were first heated to 1300 °C in Ar (4×10^{-5} mbar) and then finally heated between 1100 °C–1150 °C for 15 min in UHV. Regardless of growth protocols, the ribbons were all ballistic conductors.

It is important to note that not all samples in [58] were grown on 6H-SiC. While all the local 2- and 4-probe ballistic transport measurements were only done on 6H-SiC ribbons, other gated transport measurements in that work were all grown from 4H-SiC and their edge type was not clearly identified. The reference given for those other ribbons, Ming [62], either did not specify the ribbon facet or, in some cases, used curved ribbons that contain a mixture of AC and ZZ edges. For this section of the review, we will therefore focus on the results from Baringhaus *et al* [50] where the ribbon edge type provenance is clearly noted.

In addition to the ZZ-ribbon work of Baringhaus *et al* [36, 50, 58] and Aprojanz *et al* [59], recent work by a Georgia Tech group has confirmed that metallic graphene grows on the 6H polytype [49]. This is demonstrated in figure 1.12(e). Dirac cones from $\{11\bar{2} \sim 22\}_{6H}$ planes are found in ARPES without any passivation.

The Georgia Tech group's 6H ZZ-sidewall graphene is grown by the CSS process with the same recipes as AC-ribbons given in section 1.2.3. It was found that the best growth temperature was 1535 °C for 70 s. Heating to 1548 °C for 90 s still grew sidewall graphene but the long range order was not as good. Heating above 1564 °C for 70 s resulted in no measurable ZZ-graphene Dirac cones as measured by ARPES. The lack of Dirac cones was presumably due to step melting.

Baringhaus *et al* [36] have showed that the starting trench depth affects the sidewall facet structure when grown by the FTF method on 6H SiC. For trench depths greater than 20 nm, they see ~50° facet angles, i.e. roughly a $(11\bar{2}8)$ facet. This facet is very close to the free energy minimum corresponding to $\{11\bar{2}9\}$ planes expected for bare SiC growth (see the $F(\theta)$ plot in figure 1.4(b)). Baringhaus *et al* [36] report that ballistic transport is not seen in the large facet angle ribbons. It is possible that graphene did not grow on these steps, which would explain why the facet angle is the same as predicted for bare SiC growth. EFM shows that these deeper steps break into multiple facets with multiple parallel ZZ-ribbons on the side walls (see figures 1.14(a) and (b)). For shallower 20 nm deep trenches, the sides are smoother according to EFM (see figures 1.14(c) and (d)). However, STM data from the same group show that the sub-20 nm steps still have undulations with an averaged facet angle of ~28° (see figure 1.15(a)). 6H sidewall graphene grown from shallow trenches using the CSS method show a similar nano-facet structure in STM with a slightly smaller facet angle as measured by STM (see figure 1.15(e)). ARPES from the CSS 6H ribbons show clear facet Dirac cones as demonstrated in figure 1.12(e). The facet angle from ARPES was measured to be $\theta_F \sim 24°$ corresponding to the $\{11\bar{2} \sim 22\}_{6H}$ planes. We point out that the difference in growth temperatures between Baringhaus *et al* [36] and Miettinen *et al* [49] are related to details of the different furnace designs. These difference may be related to why 50° facet angles were not found in the CSS furnace growth samples.

The STS spectral data from different positions on the facet wall of the FTF grown ZZ-ribbons in figure 1.15(a) are shown in figures 1.15(b) and (c) [50]. They show a

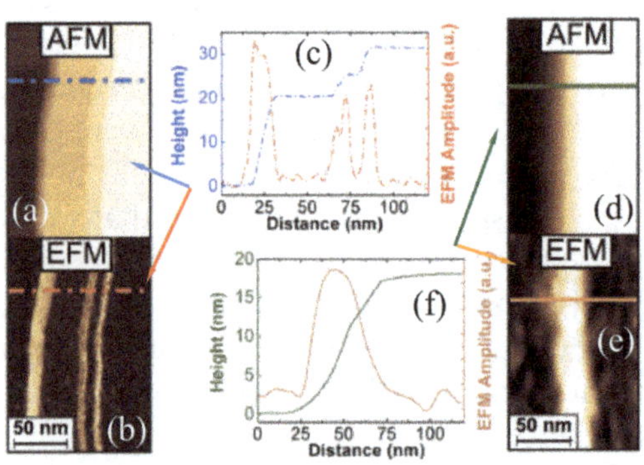

Figure 1.14. (a), (b) AFM and EFM images of a 30 nm deep step after graphene FTF growth. (c) Corresponding line scans through the AFM/EFM images in (a) and (b). (d), (e) AFM and EFM images of a 20 nm deep step after FTF graphene growth. (f) Corresponding line scans to the AFM/EFM images shown in (c). Reprinted from [36], with the permission of AIP Publishing.

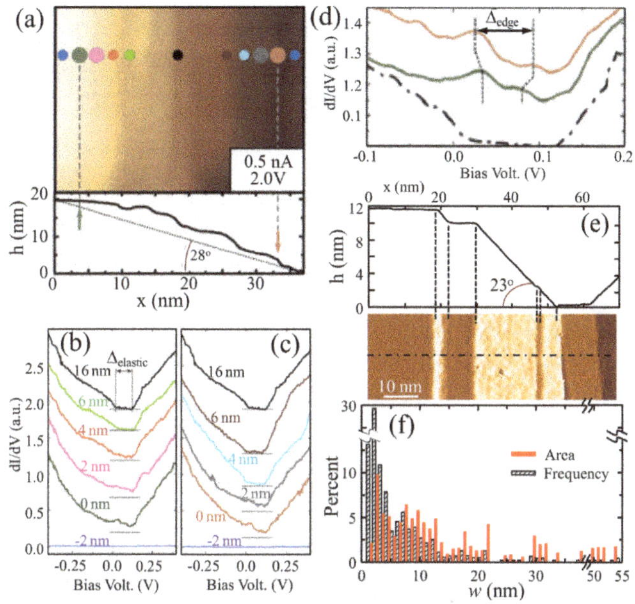

Figure 1.15. (a) STM image of a 20 nm deep 6H ZZ-graphene sidewall ribbon grown by the FTF method. Reproduced from [50]. IOP Publishing Ltd. All rights reserved. (b), (c) dI/dV spectra taken at different positions along the ribbon in (a). The color coded dots in (a) correspond to the color code in (b) and (c). Spectra are shifted for better visibility. (d) Close up of the dI/dV spectra near the top and bottom of the ribbon in (a). The dashed-dotted line denotes the spectrum taken in the center of the ribbon showing the alignment of the valence band, conduction band, and the edge-states. (e) Partial dI/dV image of a 34 nm deep 6H ZZ-graphene sidewall ribbon grown by the CSS method from [49]. (f) Histogram of 250 facet widths plotted by both relative frequency and areal coverage. Reprinted with permission from [49], Copyright (2019) by the American Physical Society.

gradual change from a gapped DOS at the center of the ribbon to a pair of states near E_F at both the top and bottom of the ribbons. The small gap of 120–130 meV in figures 1.15(b) and (c) is attributed to the onset of inelastic tunneling from the tip for energies $|E| > 65$ meV [65]. The states at the top and bottom of the ribbons were purported to be the expected edge states for ZZ-ribbons [3, 4]. While they must be related to edge states, there are differences between these states and ARPES measured states on similar 6H ZZ-ribbons. To show this, we must look at the details of CSS grown 6H-ribbon's structure and electronic properties.

Unlike AC-steps where a single $(1\bar{1}07)$ facet covers ~70% of the step area [48], 6H ZZ-steps have a complicated facet structure [49, 50, 59]. The ZZ-steps consist of many $\{11\bar{2} \sim 22\}_{6H}$-(0001) plane pairs (see figures 1.15(a) and (e)). FTF grown ribbons have essentially the same undulating facet structure as the CSS ribbons [59]. The $\{11\bar{2} \sim 22\}_{6H}$ facets have a broad width distribution, $N(w_f)$, as shown in figure 1.15(f). The histogram of $N(w_f)$ gives an average facet width of $\overline{w}_f \sim 6 \pm 8$ nm with a high number of 1–2 nm facets. Miettinen *et al* [49] showed that the facet width distribution, $N(w_f)$, is very similar to the ribbon width distribution, $N(W)$, where W is the ZZ-ribbon width. They first noted that STS maps, like those in

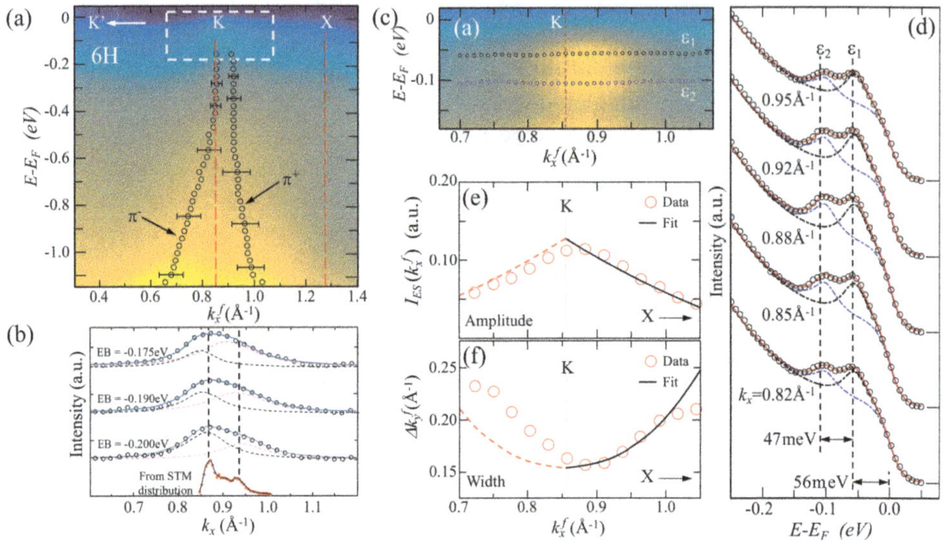

Figure 1.16. ARPES measures bands of 6H ZZ-sidewall ribbons (T = 100 K). (a) Bands along the $K - K' - \Gamma$ direction measured in the plane of the ribbon (k_x^f parallel to the ribbons edges). The peak position of the bands are marked by circles from MDC fits. (b) MDCs near $E_B \sim -0.2$ eV around the K-point showing the asymmetry in the valance bands where the π^- and π^+ bands' have merged. Red circles are the calculated asymmetry using the STM width distribution in figure 1.15(f). (c) EDC around E_F (from the dashed region in (a)) showing two weakly dispersing bands, ϵ_1 and ϵ_2, with a separation of $\Delta° = 47$ meV (circles ∘ mark the peak positions from EDCs fits). (d) Sample EDCs near E_F in (c) (∘ are data). A two Lorentzian (dashed black and blue lines) plus background and Fermi–Dirac cut off fit is shown (red solid line). (e), (f) The ϵ_1 and ϵ_2 average intensity and Δk_y^f width, respectively, versus k_x^f. Solid and dashed lines are fits using a TB model. Reprinted with permission from [49], Copyright (2019) by the American Physical Society.

figure 1.15(e), show that the facet planes are bright compared to the (0001) nano-terraces. This indicates that there is a discontinuity in the electronic structure of the graphene on the facet and the semiconducting graphene that is known to grow on the nano-terraces [48]. They argued that the metallic facet graphene either terminates into the SiC(0001) surface or transitions into a semiconducting form of graphene on the (0001) nano surface. In either case, the metallic ribbon width is proportional to the facet width. Detailed ARPES data support the same conclusion.

Figure 1.16(a) shows the 6H ribbons' band intensity for k_x^f along the $K'K\Gamma'$ direction of the $(11\bar{2} \sim 22)$ facet plane. There are two unique features of the bands that support the assertion that $w_f = W$: the π^- and π^+ bands are asymmetric and appear split for BE < −0.4 eV. The apparent splitting is indicative of an asymmetry induced in the k_x^f position of the π-band's valance band maximum, VB_m, in an area-averaged ARPES measurement when there is a distribution of ribbons widths containing a large number of sub 5 nm ribbons. In both tight binding (TB) and *ab initio* models, the k_x position of VB_m is a function of ribbon width [6, 66]. Miettinen *et al* [49] noted that the TB model gives an analytic expression for the approximate position of VB_m, k_x^M. To a very good approximation, $k_x^M \simeq k_c$ where k_c is the critical momentum given by [66]:

$$k_c = \frac{2}{a}\arccos\left[\frac{1}{2}\frac{W}{W+d}\right]. \tag{1.1}$$

$d = 2.132$ Å is the spacing between ZZ chains. Ribbons with $W \sim d$ have VB_m shifted to higher k_x. Figure 1.16(b) compares the measured asymmetry in the VB_m with the calculated k_c position from equation (1.1) using the experimental frequency weighted facet width distribution in figure 1.15(f). The agreement is remarkably good, supporting the independent ribbon model and that $W \sim w_f$. Similar estimates of the Δk_x broadening of the π-bands along with estimates of the mean BE and energy width of the VB_m (see figure 1.13(b)) can all be explained in terms of the STM measured ribbon width distribution [49].

Miettinen *et al* [49] also found that the 6H ZZ-ribbons had two nondispersing states ϵ_1 and ϵ_2, near E_F that they associated with ZZ edge states. The flat states are shown in figure 1.16(c) and in the EDCs near K in figure 1.16(d). These figures show that the BE of the two states are: $\epsilon_1 = -56$ and $\epsilon_2 = -103$ meV. Their energy width is observed to be essentially equal to the thermal broadening caused by the $T = 100$ K sample, $\Delta E = 58$ meV. However, this should not be the case if the states are associated with extended states in ribbons with a broad width distribution. Given the measured $N(W)$, such extended states would cause a W-dependent energy broadening. This is not observed. Based on the measured $N(w_f)$, the ARPES would measure an order of magnitude larger ΔE. Miettinen *et al* [49] went on to show that their k_y-width and intensity as a function of k_x are also consistent with TB predictions for ZZ-edge states (see figures 1.16(e) and (f)).

However, as first suggested by Baringhaus *et al* [50], the two states are not consistent with any model of symmetrically terminated edges. All such models predict a strong k_x-dispersion. Furthermore, symmetric models also predict that the energy of the edge state is a strong function of ribbon width [5, 6, 67–69], which is inconsistent with the small energy broadening of the two states discussed above. Instead the observed states are more consistent with asymmetric edge terminations models [70, 71]. As an example, sp^2 termination on one edge and sp^3 on the other, has been shown to give rise to nearly flat bands through the entire 1D BZ whose energies are essentially independent of the ribbon width [71].

How the ARPES-measured edge states are related to the STS-measured states remains unclear. All of the STS states (top and bottom edge) are above E_F, making them p-doped. In ARPES, the states are all n-doped. Although Baringhaus *et al* [50] claim that there are no tip effects in their STS measurements of 6H ZZ-ribbons, the difference in doping between ARPES and STS on the same system strongly implies STS tip-induced doping near E_F. Furthermore, the ARPES findings show that the facet walls are made up of multiple disconnected parallel ribbons. If the STS states were edge states, they should have been visible in many places along the facet, not just at the top and bottom of the steps. Indeed the STS states appear to persist more than 2 nm past the edge, which is an order of magnitude farther than the $1/e$ length of a ~0.3 nm expected from theoretical estimates of the edge state intensity decay [6]. It could be that resolution limitations in the STS measurements cause the states intensity to be smaller when measured on the tilted facet surface. Regardless of the details, both STS and ARPES measurements do conclude that the states are due to asymmetric edge terminations.

1.3 Summary and future outlook

The growth of EG-ribbons has made significant progress. It is now possible to grow graphene ribbons with either ZZ- or AC-edges at pre-selected locations. While the growth conditions of AC and ZZ EG ribbons are similar, their structures are not. Figures 1.17(a) and (b) compare the general structural differences between AC- and ZZ-edge sidewall graphene. After graphene growth, the 4H SiC AC-steps consist of

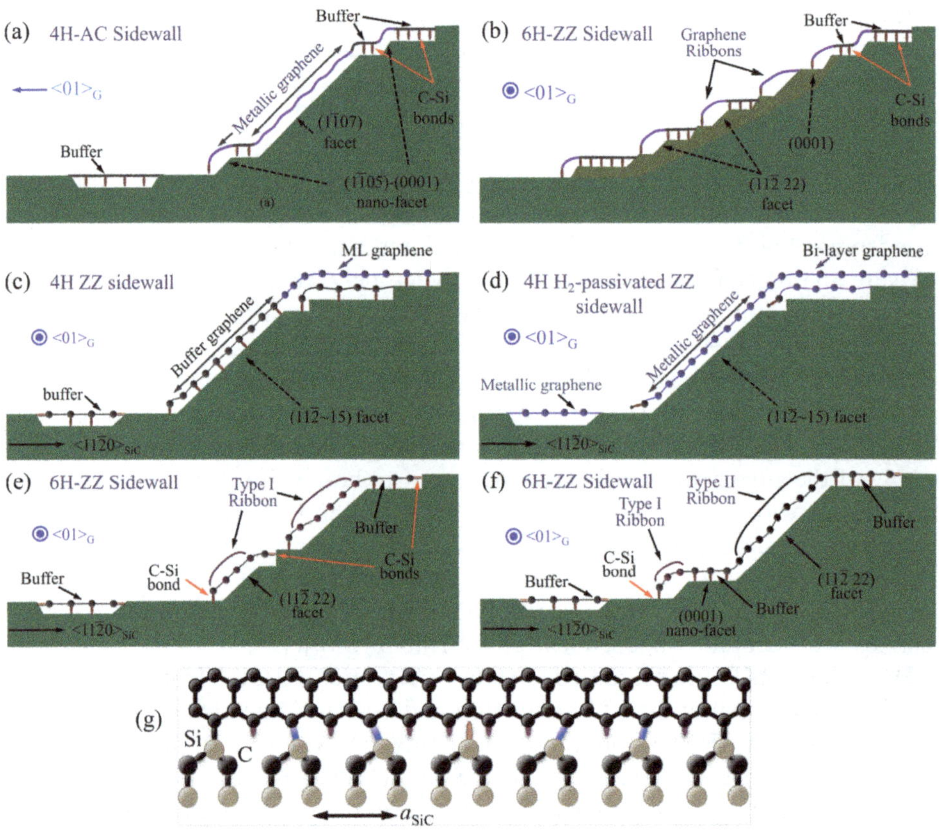

Figure 1.17. Schematic models of graphitized sidewalls for different orientations and SiC polytypes. Generic models for the structure of metallic graphene are shown in (a) for 4H AC-ribbons and (b) for 6H ZZ-ribbons. 4H ZZ-ribbons models are shown in (c) and (d). (c) The as-grown graphene bonded to $\{11\bar{2}15\}$ sidewalls. (d) The metallic ribbons after the structure in (c) has been H_2-passivated. Free edges are potentially H-terminated. (e), (f) The different edge-bonding types for 6H ZZ-ribbons. (g) Bonding geometry of a ZZ-edge ribbon to a commensurate bulk terminated $(6\sqrt{3} \times 6\sqrt{3})_{SiC}R30°$ SiC(0001) surface.

a predominantly single $\{1\bar{1}07\}$ facet bordered by a few $\{1\bar{1}05\}$-(0001) pairs on either side. The graphene on both the $\{1\bar{1}07\}$ and $\{1\bar{1}05\}$ facets is metallic. The fact that the $\{1\bar{1}05\}$ graphene remains metallic even though its width is very narrow ($\sim 2 - 3$ nm) is surprising. This suggests that the termination into semiconducting buffer graphene does not present a boundary capable to open a finite size gap. The 6H SiC ZZ-ribbons form a set of multiple parallel narrow ribbons that are generally less than 5 nm wide. The ribbons appear electronically isolated from each other and demonstrate edge state bands.

ZZ-ribbons grow differently on 4H and 6H SiC. Figure 1.17(c) shows that the ZZ-edge graphene on 4H SiC is bonded to the SiC $(11\bar{2}15)_{4H}$ facets. No electronically isolated ZZ edge graphene layer appears to grow on the facet. Instead, there is evidence that a narrow ZZ-edge electronic graphene ribbon grows on the trench tops at the step edge in the form of a ML sheet above buffer graphene (see figure 1.17(c)). When the as-grown 4H ZZ-ribbon is H_2-passivated, the graphene-substrate Si bonds are broken and the ribbons become electronically isolated from the SiC facet (see figure 1.17(d)). This makes the passivated ribbons metallic. We suppose that the bottom edge of the ribbon, where it attached to the SiC(0001), is also terminated by hydrogen-C bonds. ARPES shows that the graphene on 4H SiC is only comprised of at most a few ribbons over the entire facet. In other words, ribbon width, and thus its band structure, is determined by the original etched step height. Since the ribbons are wide, any band gap must be very small. The passivated ribbons also show no evidence of a finite size gap or n-doped edge states. Indeed, STS dI/dV measurements show that the H_2-passivated 4H ZZ-ribbons are metallic while 2-point transport measurements find them to be normal diffusive conductors rather than ballistic (see figure 1.18) [49].

On 6H-SiC, the narrow ZZ-ribbons terminate into the semiconducting buffer graphene layer on the macroscopic (0001) surface through sp^2 C-C bonds [48]. At the step bottom, the ribbon terminates by either C-Si sp^3 bonds to the substrate SiC (type I termination in figure 1.17(e)) or by an intermediate sp^2 C-C bond to buffer

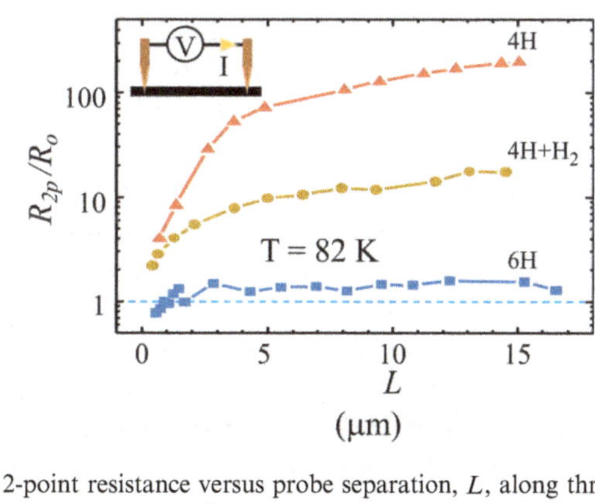

Figure 1.18. Normalized 2-point resistance versus probe separation, L, along three different ZZ-ribbon: 4H, 4H+H-passivated and 6H ribbons ($R_o = h/e^2$). Reprinted with permission from [49], Copyright (2019) by the American Physical Society.

graphene on the (0001) nano-facets (type II in figure 1.17(f)). The asymmetric type I termination is more complicated than figure 1.17(e) indicates. While the ribbon-buffer edge at the top of the step is commensurate and ordered, the C-Si sp^3 edge is incommensurate with the SiC (see figure 1.17(g)) [72]. The aperiodic C-Si sp^3 bonding leads to >60% bond defects with the edge-carbon either unbonded or re-hybridized in some complicated way. Regardless of whether or not the ribbons are type I or II, their terminations are asymmetric. Asymmetric ZZ-edge terminations are known to lead to flatter band over the entire zone [71]. Baringhaus *et al* [50] suggested that the graphene-SiC termination may act in a similar way to the changes predicted to occur when line defects run periodically parallel with the ribbon's ZZ-edges [73]. The defects lead to weakly dispersing bands and lift the degeneracy in the ferromagnetic states [73]. Regardless of the details of the asymmetric C-Si sp^3 edges, the narrow energy widths and dispersionless character of the observed ϵ_1 and ϵ_2 bands are consistent with the edge states from asymmetric edge terminations in the sidewall SiC system.

The structural differences between 4H and 6H ZZ-ribbons are directly related to their transport properties. Figure 1.18 compares the 2-probe resistance as a function of probe separation along a sidewall. As already mentioned, the wide 4H H_2-passivated ribbons are diffusive conductors. For the narrow 6H ZZ-sidewall graphene, the ribbon resistance is independent of probe spacing up to at least 16 μm. Furthermore, the resistance value is $R_o = h/e^2$, meaning that the 6H ZZ-ribbons are ballistic conductors [49, 58, 59]. STS measurements on unpassivated 4H ZZ-ribbons also indicate that they are semiconducting [74]. This would be the expected result given that the ribbons are bonded to the SiC through a very similar interaction that causes the buffer graphene layer on the (0001) surface to be semiconducting.

The differences between transport on 4H and 6H ZZ-ribbons is puzzling in terms of previously published results. Fixed geometry transport measurements on 4H ZZ-sidewall ribbons were used to explain local ballistic transport in 6H ZZ-ribbon [58]. The large asymmetry in the 4H-ribbon conductance as a function of gate voltage was attributed to the $n = 0$ subbands (edge state) being the conducting channel with the $n \geqslant 1$ subbands being gapped. How can fixed transport measurements on essentially semiconducting 4H-sidewall graphene explain local probe ballistic transport on 6H-sidewall ribbons? There are several possibilities. First, the local probe transport purposefully studied the graphene on the sidewall. As discussed in section 1.2.3, there is a metallic ZZ-ribbon on the (0001) surface at the step edge. It is possible that the fixed transport gate, which covers ribbons on both (0001) and facet planes, measure transport from the (0001) ribbons; shorting out the semiconducting graphene on the sidewalls. A second possibility is that the formation of the gate oxide in the fixed transport studies delaminates the semiconducting 4H ZZ-ribbon. There is unpublished work suggesting that fixed gate oxides can change the semiconducting buffer graphene's band structure [74]. Finally, the original 4H ZZ-ribbon work [62] presented in [58] was done on natural steps, not patterned steps. It is entirely possible that the graphene on the shallow natural 4H-steps is not attached to the SiC facets and is thus conducting.

These questions and their potential explanations for the transport properties of ZZ-ribbons underscore the need for more work on sidewall graphene ribbons. In particular, studies of shallow 4H ZZ-trenches need to be carried out. If sidewall ribbons are to be part of a new emerging carbon electronic revolution, the relationship between ribbon geometry and transport needs to be carefully studied.

References

[1] Berger C *et al* 2004 *J. Phys. Chem.* B **108** 19912
[2] Berger C *et al* 2006 *Science* **312** 1191
[3] Fujita M, Wakabayashi K, Nakada K and Kusakabe K 1996 *J. Phys. Soc. Jpn.* **65** 1920
[4] Wakabayashi K, Fujita M, Ajiki H and Sigrist M 1999 *Phys. Rev.* B **59** 8271
[5] Son Y-W, Cohen M L and Louie S G 2006 *Phys. Rev. Lett.* **97** 216803
[6] Yang L, Park C-H, Son Y-W, Cohen M L and Louie S G 2007 *Phys. Rev. Lett.* **99** 186801
[7] Takane Y and Wakabayashi K 2008 *J. Phys. Soc. Jpn.* **77** 054702
[8] Mucciolo E R, Castro Neto A H and Lewenkopf C H 2009 *Phys. Rev.* B **79** 075407
[9] Norimatsu W and Kusunoki M 2010 *Physica* E **42** 691
[10] Han M Y, Özyilmaz B, Zhang Y and Kim P 2007 *Phys. Rev. Lett.* **98** 206805
[11] Todd K, Chou H-T, Amasha S and Goldhaber-Gordon D 2009 *Nano Lett.* **9** 416
[12] Ritter K A and Lyding J W 2009 *Nat. Mater.* **8** 235
[13] Han M Y, Brant J C and Kim P 2010 *Phys. Rev. Lett.* **104** 056801
[14] Sols F, Guinea F and Neto A H C 2007 *Phys. Rev. Lett.* **99** 166803
[15] Jiao L, Zhang L, Wang X, Diankov G and Dai H 2009 *Nature* **458** 877
[16] Magda G Z, Jin X, Hagymasi I, Vancsó P, Osváth Z, Nemes-Incze P, Hwang C, Biró L and Tapasztó L 2014 *Nature* **514** 608
[17] Cai J *et al* 2010 *Nature* **466** 470
[18] Tao C *et al* 2011 *Nat. Phys.* **7** 616
[19] Ruffieux P *et al* 2012 *ACS Nano* **6** 6930
[20] Linden S *et al* 2012 *Phys. Rev. Lett.* **108** 216801
[21] Ruffieux P *et al* 2016 *Nature* **531** 489
[22] Liu J, Li B-W, Tan Y-Z, Giannakopoulos A, Sanchez-Sanchez C, Beljonne D, Ruffieux P, Fasel R, Feng X and Müllen K 2015 *J. Am. Chem. Soc.* **137** 6097
[23] Sprinkle M, Ruan M, Hu Y, Hankinson J, Rubio-Roy M, Zhang B, Wu X, Berger C and de Heer W A 2010 *Nat. Nano.* **5** 727
[24] Kajiwara T, Nakamori Y, Visikovskiy A, Iimori T, Komori F, Nakatsuji K, Mase K and Tanaka S 2013 *Phys. Rev.* B **87** 121407
[25] Bommel A J V, Crombeen J E and Tooren A V 1975 *Surf. Sci.* **48** 463
[26] Hass J, de Heer W A and Conrad E H 2008 *J. Phys. Condens. Matter* **20** 323202
[27] Hu Y, Ruan M, Guo Z, Dong R, Palmer J, Hankinson J, Berger C and de Heer W A 2012 *J. Phys. D: Appl. Phys.* **45** 154010
[28] de Heer W A, Berger C, Ruan M, Sprinkle M, Li X, Hu Y, Zhang B, Hankinson J and Conrad E 2011 *Proc. Natl Acad. Sci.* **108** 16900
[29] Robinson J, Weng X, Trumbull K, Cavalero R, Wetherington M, Frantz E, LaBella M, Hughes Z, Fanton M and Snyder D 2010 *ACS Nano* **4** 153
[30] Hicks J, Shepperd K, Wang F and Conrad E H 2012 *J. Phys. D: Appl. Phys.* **45** 154002
[31] Nevius M S 2016 Improved growth, ordering and characterization of sidewall epitaxial graphene nanoribbons *PhD Thesis* Georgia Institute of Technology

[32] Nevius M S, Wang F, Mathieu C, Barrett N, Sala A, Mentes T O, Locatelli A and Conrad E H 2014 *Nano. Lett.* **14** 6080
[33] Powell J A, Neudeck P G, Trunek A J, Beheim G M, Matus L G, Jr R W H and Keys L J 2000 *Appl. Phys. Lett.* **77** 1449
[34] Claire B *et al* 2016 Epitaxial graphene on SiC: 2D sheets, selective growth and nanoribbons *Graphene Growth on Semiconductors* ed N Motta, F Iacopi and C Coletti (Singapore: Pan Stanford), pp 181–204
[35] Yu X, Hwang C, Jozwiak C, Kohl A, Schmid A and Lanzara A 2011 *J. Electron. Spectrosc. Relat. Phenom.* **184** 100
[36] Baringhaus J, Aprojanz J, Wiegand J, Laube D, Halbauer M, Hübner J, Oestreich M and Tegenkamp C 2015 *Appl. Phys. Lett.* **106** 043109
[37] Emtsev K V *et al* 2009 *Nat. Mater.* **8** 203
[38] Conrad M D 2017 Structure and properties of incommensurate and commensurate phases of graphene on SiC(0001) *PhD Thesis* Georgia Institute of Technology
[39] Emtsev K V, Speck F, Seyller T, Ley L and Riley J D 2008 *Phys. Rev.* B **77** 155303
[40] Nevius M S, Conrad M, Wang F, Celis A, Nair M N, Taleb-Ibrahimi A, Tejeda A and Conrad E H 2015 *Phys. Rev. Lett.* **115** 136802
[41] Ouerghi A, Silly M G, Marangolo M, Mathieu C, Eddrief M, Picher M, Sirotti F, El Moussaoui S and Belkhou R 2012 *ACS Nano* **6** 6075
[42] Nicotra G, Ramasse Q M, Deretzis I, La Magna A, Spinella C and Giannazzo 2013 *ACS Nano* **7** 3045
[43] Giannazzo F, Deretzis I, Nicotra G, Fisichella G, Spinella C, Roccaforte F and Magna A L 2014 *Appl. Surf. Sci.* **291** 53
[44] Giannazzo F, Deretzis I, Nicotra G, Fisichella G, Ramasse Q, Spinella C, Roccaforte F and Magna A L 2014 *J. Cryst. Growth* **393** 150
[45] Sforzini J *et al* 2015 *Phys. Rev. Lett.* **114** 106804
[46] Nordell N, Karlsson S and Konstantinov A 1999 *Mater. Sci. Eng.* B **61–2** 130
[47] Hicks J *et al* 2013 *Nat. Phys.* **9** 49
[48] Palacio I *et al* 2015 *Nano. Lett.* **15** 182
[49] Miettinen A, Nevius M S, Ko W, Kolmer M, Nair M N, Kierren B, Moreau L, Conrad E H and Tejeda A 2019 *Phys. Rev.* B **100** 045925
[50] Baringhaus J, Edler F and Tegenkamp C 2013 *J. Phys. Condens. Matter* **25** 392001
[51] Hicks J D 2013 A combined top-down/bottom-up route to fabricating graphene devices *PhD Thesis* Georgia Institute of Technology
[52] Lu C-Y, Cooper J A, Tsuji T, Chung G, Williams J R, McDonald K and Feldman L C 2003 *IEEE Trans. Electron Devices* **50** 1582
[53] Baskin Y and Meyer L 1955 *Phys. Rev.* **100** 544
[54] Shirley E L, Terminello L J, Santoni A and Himpsel F J 1995 *Phys. Rev.* B **51** 13614
[55] Bostwick A, Ohta T, McChesney J L, Emtsev K V, Seyller T, Horn K and Rotenberg E 2007 *New J. Phys.* **9** 385
[56] Knox K R, Wang S, Morgante A, Cvetko D, Locatelli A, Mentes T O, Nino M A, Kim P and Osgood R M 2008 *Phys. Rev.* B **78** 201408
[57] Lee D S, Riedl C, Krauss B, von Klitzing K, Starke U and Smet J H 2008 *Nano. Lett.* **8** 4320
[58] Baringhaus J *et al* 2014 *Nature* **506** 349
[59] Aprojanz J, Power S R, Bampoulis P, Roche S, Jauho A-P, Zandvliet H J W, Zakharov A A and Tegenkamp C 2018 *Nat. Commun.* **9** 4426

[60] Nakajima A, Yokoya H, Furukawa Y and Yonezu H 2005 *J. Appl. Phys.* **97**
[61] Nie S, Lee C D, Feenstra R M, Ke Y, Devaty R P, Choyke W J, Inoki C K, Kuan T S and Gu G 2008 *Surf. Sci.* **602** 2936
[62] Ming F 2011 Structure epitaxial graphene for electronics *PhD Thesis* Georgia Institute of Technology
[63] Berger C, Conrad E and De Heer W W D H 2017 *Epigraphene: Physics of Solid Surfaces* ed G Chiarotti and P Chiaradia vol III45B (Berlin: Springer), p 727–807
[64] Conrad M, Rault J, Utsumi Y, Garreau Y, Vlad A, Coati A, Rueff J-P, Miceli P F and Conrad E H 2017 *Phys. Rev.* B **96** 195304
[65] Zhang Y, Brar V W, Wang F, Girit C, Yayon Y, Panlasigui M, Zettl A and Crommie M F 2008 *Nat. Phys.* **4** 627
[66] Wakabayashi K, Sasaki K-i, Nakanishi T and Enoki T 2010 *Sci. Technol. Adv. Mater.* **11** 54504
[67] Pisani L, Chan J A, Montanari B and Harrison N M 2007 *Phys. Rev.* B **75** 064418
[68] Jung J and MacDonald A H 2009 *Phys. Rev.* B **79** 235433
[69] Yazyev O V 2010 *Rep. Prog. Phys.* **73** 056501
[70] Lee G and Cho K 2009 *Phys. Rev.* B **79** 165440
[71] Deng X, Zhang Z, Tang G, Fan Z and Yang C 2014 *Carbon* **66** 646
[72] Conrad M *et al* 2017 *Nano Lett.* **17** 341
[73] Dutta S and Wakabayashi K 2015 *Sci. Rep.* **5** 11744
[74] Miettinen A 2019 Epitaxial zig-zag graphene ribbons; structure and electronic properties *PhD Thesis* Georgia Institute of Technology

IOP Publishing

Graphene Nanoribbons

Luis Brey, Pierre Seneor and Antonio Tejeda

Chapter 2

Bottom-up approach for the synthesis of graphene nanoribbons

Joffrey Pijeat, Jean-Sébastien Lauret and Stéphane Campidelli

2.1 Introduction

The bottom-up approach is a chemical method used for the fabrication of nanomaterials. In contrast with top-down methods [1–3], in which the material is directly carved from its bulk mater, the strategy of the bottom-up approach is based on the controlled assembly of small entities used as building blocks for the creation of larger and structurally well-defined nanomaterials. Applying the bottom-up approach for the fabrication of graphene nanoribbons (GNRs) allows controlling at the atomic level of both the geometry of the materials including edges state, defects, size, etc, and the chemical composition which permits notably controlling doping parameters, such as the ratio, type and position of dopants. The access to structurally well-defined GNRs allows scientists to better understand, predict, tailor and improve the electronic properties of materials. Historically, the fabrication of GNRs was first achieved in solution by using organic chemistry strategies. Later on, a surface assisted protocol was developed giving access to a large variety of GNRs that have been observed and characterized using scanning tunneling microscopy techniques. Finally, by using band gap engineering, examples of atomically well-defined heterojunctions emerged. These junctions were composed by at least two GNRs segments exhibiting different widths, topologies or doping.

2.2 Solution-mediated synthesis graphene nanoribbons

2.2.1 Synthesis of early narrow graphene nanoribbons

From a chemical point of view, the structure of a GNR can be seen as multiple fused sp^2 carbon rings forming a flat and electronically delocalized polymer. Therefore, although not recognized in those days as GNRs, examples of fully conjugated polymers can be found in the 1970s and 1990s (figure 2.1). These ribbon-like

doi:10.1088/978-0-7503-1701-6ch2

© IOP Publishing Ltd 2020

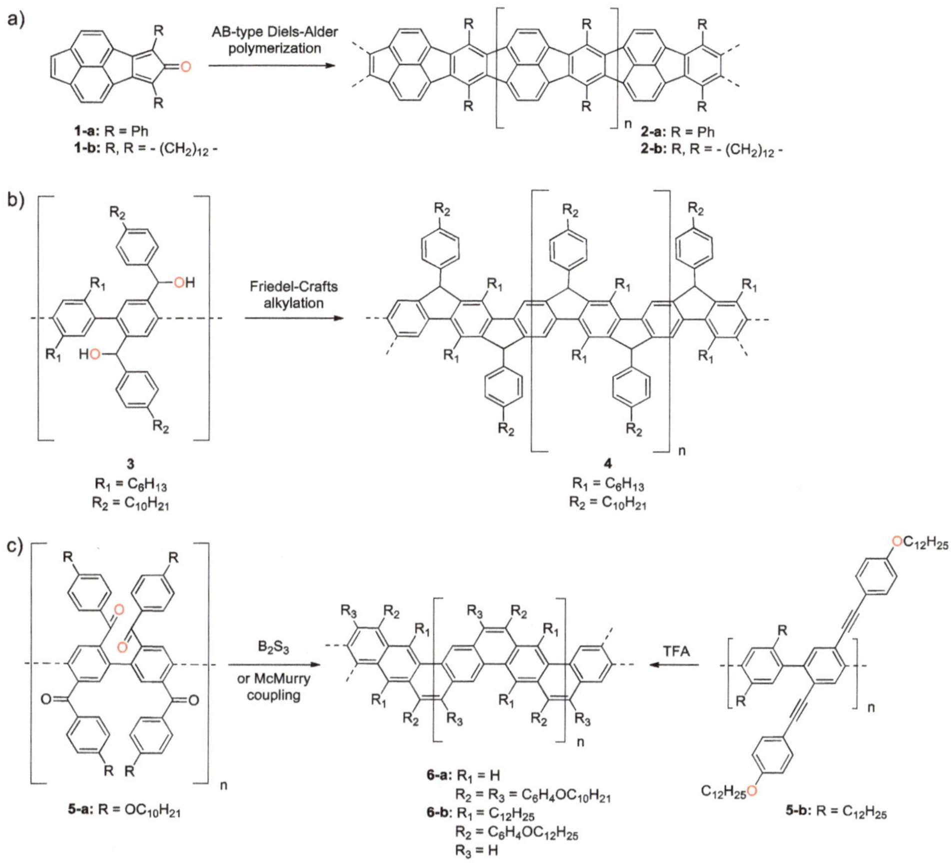

Figure 2.1. Early ladder type polymers found in literature in the 1970s and the 1990s. These polymers were synthesized in solution via different synthetic strategies: (a) AB-type Diels-Alder polymerization [4]; (b) Friedel-Craft alkylation [6, 7]; (c) carbonyl olefination [8, 9], McMurry coupling or intramolecular electrophilic substitution under acid condition [10].

polymers can nowadays be seen as the 'narrowest GNRs' and were synthesized via solution-mediated protocols.

The first attempt for the synthesis of pure carbon based ladder-type polymer **2-a** was achieved by Stille *et al* [4] via AB-type Diels-Alder polymerization from monomer **1-a** containing both the diene and dienophile on the same molecule. However, the low solubility of the resulting product, due to strong π-π interactions, made its characterization rough at that time. The solubility issue was overcome in 1994 by Löffler *et al* [5] by incorporating flexible alkyl loops on the monomer **1-b**, making the characterization of polymer **2-b** possible and confirming the structure. In 1991, Scherf and Müllen reported the synthesis of ladder type polymer **4** from precursor **3** via intramolecular Friedel-Craft alkylation [6, 7]. Polymer **4** was soluble in organic solvents thanks to the presence of alkyl chains at the periphery of the ribbon. Other alternative synthetic strategies emerged in the 1990s using carbonyl olefination [8, 9], McMurry coupling or intramolecular electrophilic substitution to form angular polyacene **6-a** and **6-b** [10].

Although all of these ladder type polymers can be seen as the 'narrowest' GNRs, their narrow widths do not provide the high charge carrier mobility expected for GNRs and hence the target electronic properties of GNRs cannot be reached by such systems. Therefore, still by applying a bottom-up approach, the necessity to form larger GNRs fostered scientists to propose other synthetic strategies.

2.2.2 Solution mediated synthesis of more extended GNRs

In the 2000s, the group of Müllen developed a strategy for the formation of extended armchair-edge GNRs (A-GNRs) [11, 12]. The first example, GNR **7** ($N = 9$), was obtained from cyclodehydrogenation of precursor **8** synthesized via A_2B_2 polymerization using Diels-Alder reactions, between 1, 4-bis(2, 4, 5-triphenylcyclopentadienone-3-yl)benzene **9** as *bis*-diene and 1, 4-diethynylbenzene **10** as *bis*-dienophile (figure 2.2(a)) [13]. Size exclusion chromatography (SEC) performed on polyphenylene **8** indicated a weight average molar mass (M_w) ranging from 12 000 to 120 000 g mol^{-1}. However, the asymmetry of starting monomer **9** caused structural isomerization during the Diels-Alder reaction, which leads to a certain number of regioisomers, further increased after cyclodehydrogenation of **8** because of the free rotations of certain C–C bonds. Therefore, GNR **7** was formed in a mixture of three regioisomeric repeating units *x*, *y*, *z* and presented a non-linear structure with irregular edges and was insoluble in common solvents.

In 2003, the 1,4-diethynylbenzene **10** was replaced by an extended analogue, a 1,4-diethynyl-2,5-di(4'-tert-butylphenyl)benzene **13** to fill the gaps between the repeating units and correct the homogeneity of lateral edges (figure 2.2(b)). Nevertheless, because of the shape of the monomers, GNR **12** was obtained as a mixture of randomly kinked polymers [14]. An average number molecular weight M_n of 25 000 g mol^{-1} was determined by SEC and a polydispersity index (PDI) of 2.5, which corresponded to nanoribbons with 6.5 width nm and 35 nm length according to high resolution transmission electron microscopy (HRTEM).

In 2009, still by applying a Diels-Alder strategy, a series of linear GNRs emerged with gulf-type edges. Cycloadditions were performed between monomers **14**, **15**, **16** and formed a mixture of oligophenylenes **17** from $n = 1$ to $n = 5$, which afforded short nanoribbons **18** containing up to 373 sp^2 carbon (for $n = 5$) after cyclodehydrogenation (figure 2.3(a)). The smallest ribbon **18** ($n = 1$) was comprehensively characterized but larger oligomers were insoluble in common organic solvents making their characterizations difficult [13]. In 2014, Narita *et al* succeeded in the synthesis of longer gulf type GNR **19** by oxidation of precursor **20** preliminary formed by polymerization of a AB-type monomer **21** made of a cyclopentadienone containing an ethynyl group (figure 2.3(b)) [14]. Before oxidation, the polyphenyl precursor **20** exhibited a weight average molar mass M_w of 470 000 g mol^{-1} and for GNR **19** an optical band gap of 1.88 eV was estimated. In the same year, larger GNR **22** was obtained from laterally extended pentadienone **23** using the same AB-type Diels-Alder methodology (figure 2.3(c)) [15]. The unoxidized polymer precursor

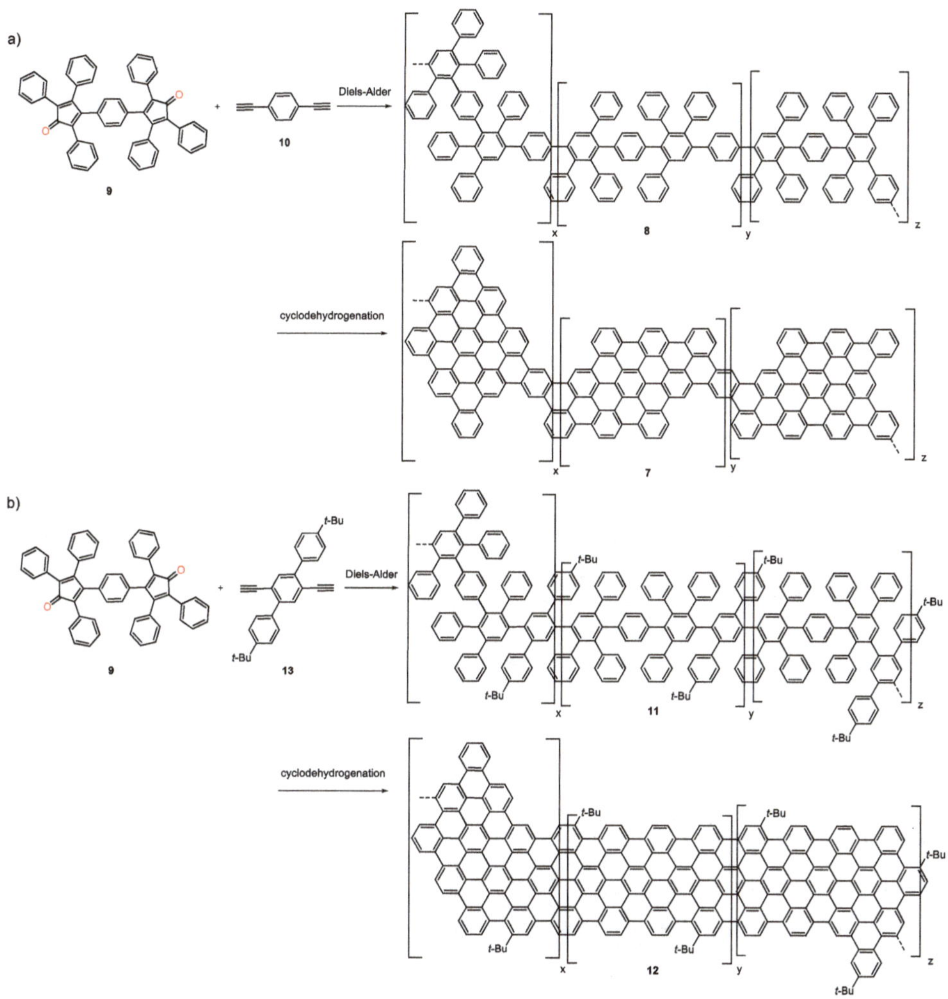

Figure 2.2. Synthetic strategy developed by the group of Müllen for the fabrication of laterally extended-A-GNRs 7 and 12 using a two steps protocol: (1) Diels-Alder polymerization, (2) cyclodehydrogenation of the subsequent polyphenylene precursor [11, 12].

24 exhibited a M_w ranging from 230 000 to 550 000 g mol^{-1} and a lower optical band gap of 1.24 eV was calculated for GNR **22**.

In 2008, Müllen *et al* proposed the fabrication of new armchair-type GNRs using Suzuki-Miyaura cross-coupling reactions. GNR **25** was synthesized by the A_2B_2 polymerization of two monomers **26** and **27** bearing two boron pinacol and iodo functional groups, respectively (figure 2.4(a)) [16]. The polyphenyl polymer **28** showed an average number molecular weight M_n of 14 000 g mol^{-1} and a PDI of 1.2. According to scanning tunnelling microscopy (STM), this M_n corresponded to an average length of 8-12 nm. This relatively short size was due to the polymerization process which might be limited by steric hindrance in the monomers. These limitations were tackled in 2016 by the group of Dong, who performed AB-type

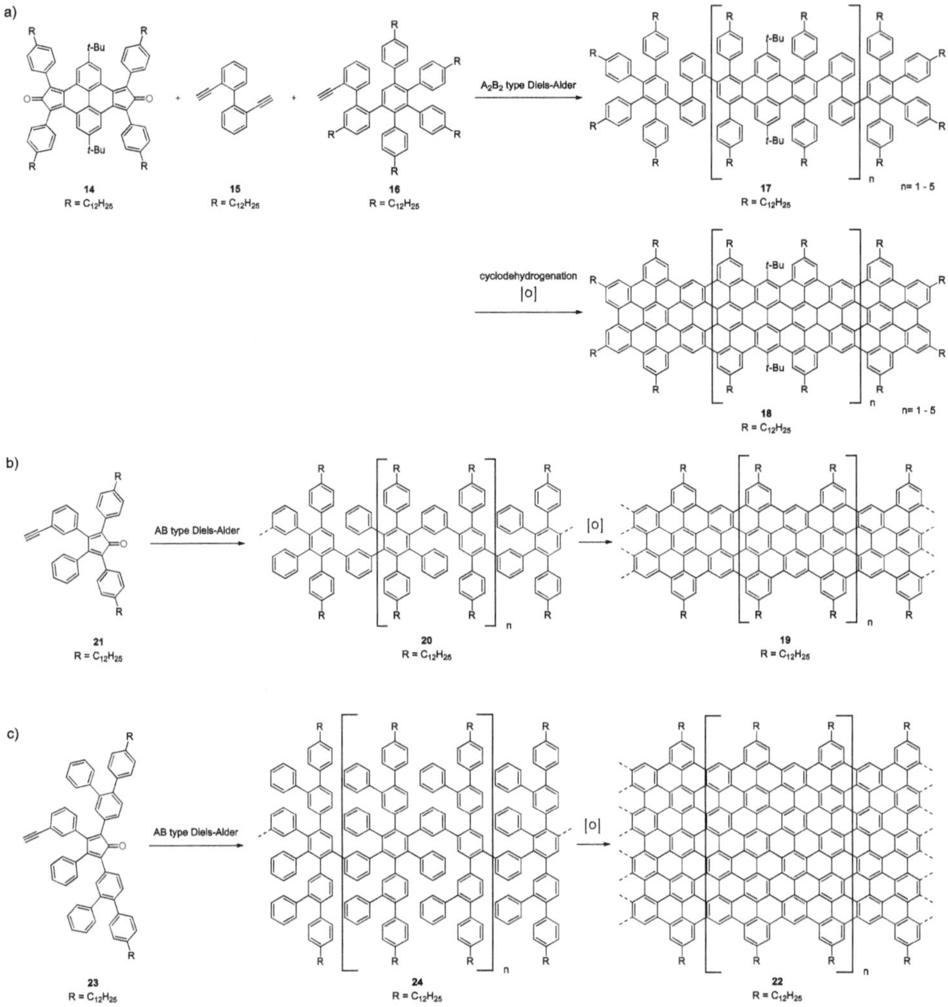

Figure 2.3. Synthetic route for gulf type-edge GNRs **18**, **19** and **22** via Diels-Alder strategy. Lateral functionalization with dodecyl chains enhanced the solubility of ribbons and facilitated their characterizations [13–15].

Suzuki-Miyaura polymerization of a unique monomer **29** containing both AB type functions (figure 2.4(b)) to form the polymer **30** which led to GNR **25** after cyclodehydrogenation [17]. Polymer **30** exhibited a M_n of 30 600 g mol^{-1} and a PDI of 1.4 and a band gap of approximately 1.1 eV was obtained by UV–vis-NIR spectroscopy and cyclic voltammetry for GNR **25**.

Another example of GNR synthesis using A_2B_2 Suzuki-Miyaura polymerization was reported in 2011 (figure 2.4(c)) [18]. The chevron-type GNR **31** was obtained by oxidation of the corresponding polymer **32** which was synthesized by polymerization of 1,2-dibromobenzene derivative **33** with 1,4-benzenediboronic acid bis(pinacol) ester **34**. Polyphenylene **32** exhibited a M_n of 9900 g mol^{-1} and a PDI of 1.4, which

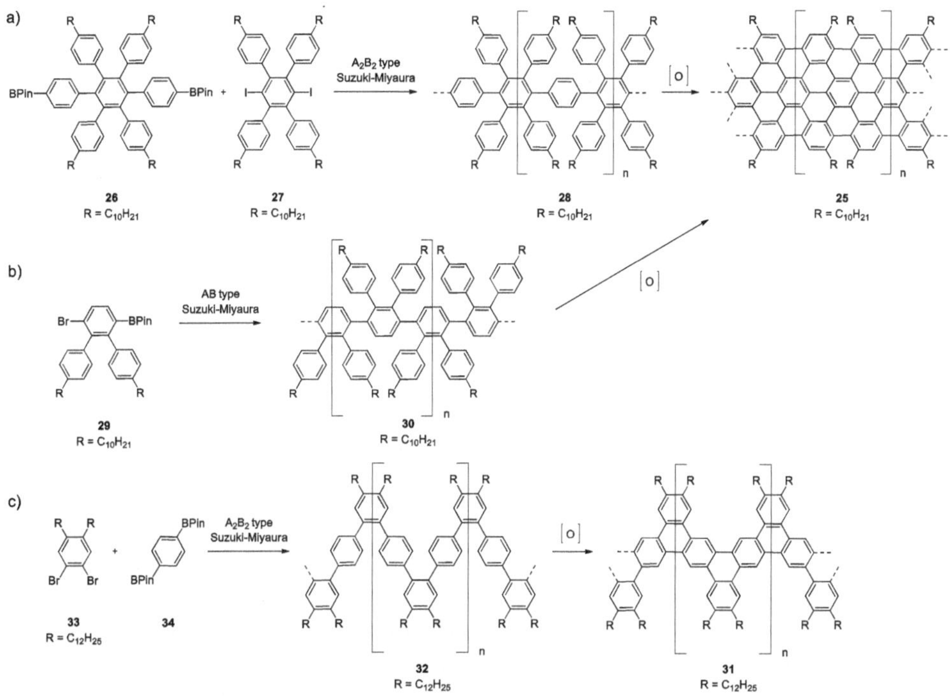

Figure 2.4. AB-type Suzuki-Miyaura cross coupling polymerization for the solution-mediated syntheses of GNRs 25 [16–18].

led to an average length of 25 nm for GNR **31** after graphitization. Thanks to the presence of dodecyl chains at the periphery, the ribbon was soluble in common solvents.

In 2016, Yang *et al* [19] developed an acetylene containing molecule to form narrow GNR **35** (figure 2.5(a)). The formation of the precursor polymer **36** was based on AB-type Suzuki-Miyaura polymerization of monomer **37**. Finally, the formation of the π-conjugated ribbons was not based on the classical Scholl reaction but was performed by benzannulation reactions of the pendant alkyne groups. The weight average molar mass and polydispersity index of **36** determined by SEC were dependent on the solvent in which the Suzuki-Miyaura polymerizations were performed and reached a M_w 37 600 g mol^{-1} and a PDI of 1.4 in THF. A similar benzannulation strategy was used by Gao *et al* [20] to synthesize GNR **38** (figure 2.5(b)). First, the backbone polymer containing acetylenic functions **39** was synthesized by A_2B_2 Sonogashira polymerization of the diethynyl derivative **40** with 1,4-diiodobenzene **41** and 4-*tert*-butyl-iodobenzene **42** (acting as polymer ends). Then the benzannulation reaction of the backbone polymer **39** with the aldehyde derivative **43** led to the precursor polymer **44** which was dehydrogenated in the presence of DDQ (2,3-dichloro-5,6-dicyano-1,4-benzoquinone) to give GNR **38**. The GNRs were tested in a series of field-effect devices; however, because of the strong aggregation state of the GNR, a band gap of ca. 0.2 eV was determined by

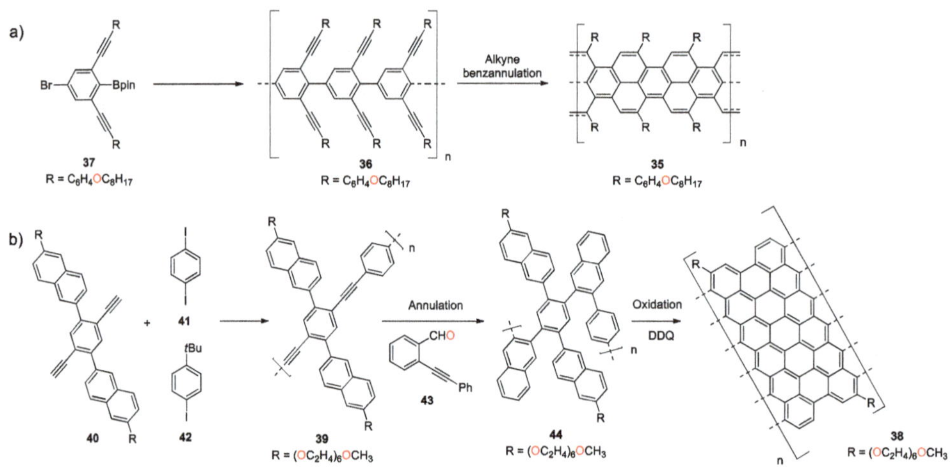

Figure 2.5. Annulation strategy developed for the conception of narrow GNRs in solution [19, 20].

electrical measurements significantly lower than the optical band gap of 1.1 eV found absorption spectroscopy.

Other transition metal catalyzed polymerizations were tested to produce GNRs. By circumventing the synthetic or stoichiometric drawbacks of the A_2B_2 or AB Diels-Alder, Suzuki-Miyaura or Sonogashira polymerizations, AA-type Yamamoto cross coupling reaction is believed to yield high molecular weight polymer with a more efficient process [21, 22]. The Yamamoto reaction in solution is somehow the image of the Ullmann coupling on the surface used for the preparation of GNRs, which will be described in the second part of this chapter.

In 2012, Müllen *et al* reported the fabrication of gulf edge GNR **45** by Yamamoto polymerization of monomer **46** bearing two halogen atoms (here chlorine) in the presence of a nickel catalyst (figure 2.6(a)) [23]. The polyphenylene polymer **47** exhibited a M_w of 52 000 g mol^{-1} and a PDI of 1.2 and an optical band gap of 1.12 eV was determined by UV–vis-NIR spectroscopy for GNR **45**. In 2014, the group of Sinitskii reported the structure of chevron-type GNR (c-GNR) **48** by polymerization in solution of monomer **49** followed by the Scholl oxidation of the resulting polymer **50** (figure 2.6(b)) [24]. *Note that GNR* **48** *is also one of the first GNR made on surface* (see section 2.3). Recently, the same group described the synthesis of c-GNR **51** which is a version of the c-GNR **48** in which a phenyl ring has been added on top of the monomer (see compound **52**, figure 2.6(c)) [25]. After Yamamoto reaction, polymer **53** showed a M_w of 30 000 g mol^{-1} and a PDI of 3.2. For the two GNRs, very similar band gaps were determined by UV–vis-NIR spectroscopy for **48** and **51** (1.6 versus 1.5 eV, respectively).

The formation of doped GNRs was also reported in solution. In 2014, the group of Sinitskii described the synthesis of chevron-like GNR **55** by Yamamoto polymerization of monomer **55** containing a pyrimidine ring followed by oxidation of polyphenylene dendrimer **56** (figure 2.7) [26]. The presence of the nitrogen in the

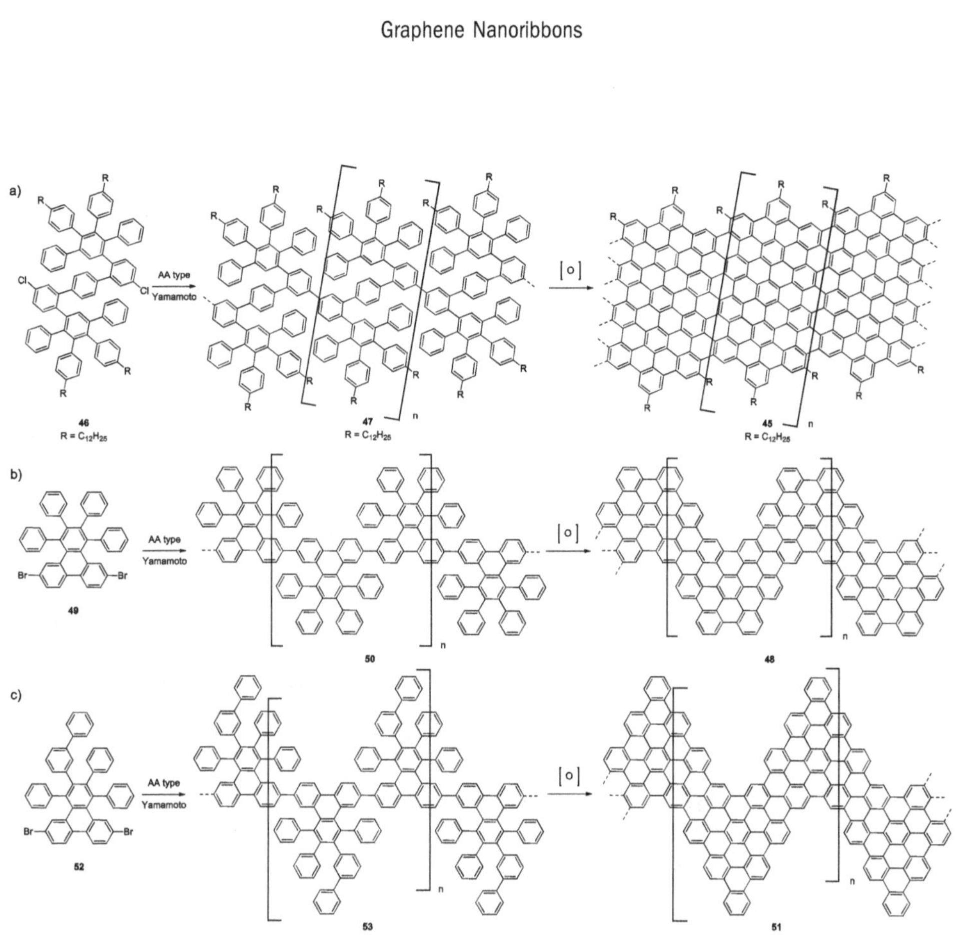

Figure 2.6. AA-type Yamamoto cross coupling polymerization for solution-mediated syntheses of GNRs **45** [23], **48** [24] and **51** [25].

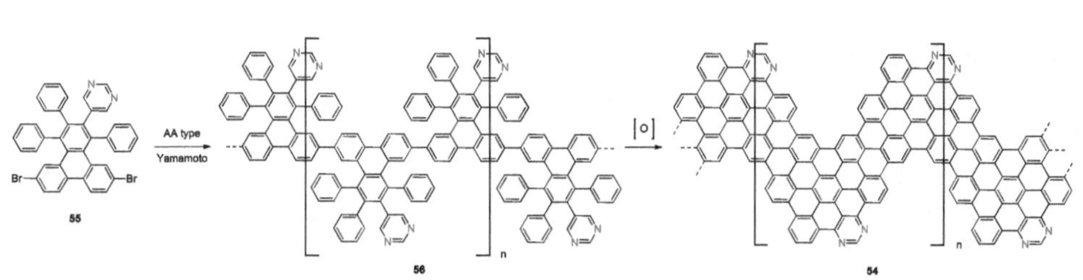

Figure 2.7. AA-type Yamamoto polymerization strategy for the conception of nitrogen-doped GNR **54** [26].

GNR was proved by x-ray photoemission (XPS) and energy-dispersive x-ray spectroscopy (EDX).

Finally, several examples of functionalization of GNR have been reported; the functionalization is performed after the polymerization step and can be performed either before or after the oxidation of the polyphenylene into GNR. The functionalization of GNRs with polymeric side chains can permit to modulate the physicochemical properties [27]. The functionalization with polycyclic aromatic hydrocarbon (PAH) allowed us to control the self-assembled structure to program

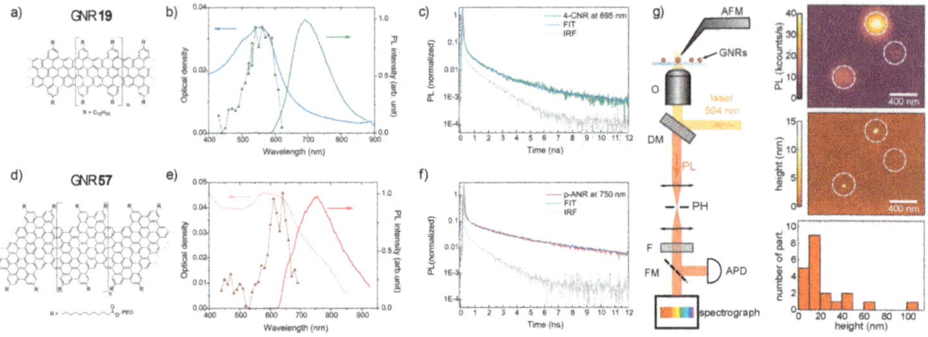

Figure 2.8. (a) Structure of GNR 19; (b) absorption (blue), photoluminescence (green) and photoluminescence excitation (green triangles) spectra of GNR 19 in sodium dodecylsulfate 2 wt% in water; (c) time-resolved photoluminescence (green), bi-exponential decay fit (blue) and IRF (grey); (d) structure of GNR 57; (e) absorption (pink), photoluminescence (red) and photoluminescence excitation (brown triangles) spectra of GNR 57 in sodium dodecylsulfate 2 wt% in water; (f) time-resolved photoluminescence (pink), bi-exponential decay fit (blue) and IRF (grey); (g) representation of the AFM/MicroPL setup with the PL and AFM images of GNR 57 deposited on a coverslip and histogram of the height of the luminescent particles on this particular sample. Reprinted from [31], Copyright (2017), with permission from Elsevier.

the supramolecular architecture [28]. It is also possible to add new magnetic properties to GNRs by functionalization with radical molecules [29].

In 2013, Müllen *et al* developed a protocol to precisely chlorinate the edges of GNRs in solution. As also demonstrated for nanoparticles of graphene, the substitution with chlorines on the edges of a GNR enhanced its solubility and decrease the optical band gap [30].

2.2.3 Optical properties of GNRs

The GNRs synthesized by solution mediated protocols have been investigated through optical experiments. In particular, GNR **19** [14] and GNR **57** [27] have been studied both in solution and in the solid state by means of photoluminescence (PL), excitation of the photoluminescence (PLE) and time resolved photoluminescence experiments (TR-PL) [31].

Figure 2.8 shows the structure of the GNRs and their optical absorption, PL and PLE spectra in suspension of sodium dodecylsulfate 2 wt % in water. The PL spectra of the GNRs (green and red curves in figures 2.8(b) and (e), respectively) exhibit broad and featureless characters. The PLE spectra (green and brown triangle curves) fit quite well the absorption spectrum ensuring that the luminescence does arise from the GNRs in suspension. Moreover, the TR-PL experiments display non mono-exponential behaviors. These observations lead to attributing the emission to excimer states in small aggregates of GNRs. This interpretation is further confirmed by experiment in the solid state. Indeed, figure 2.8(g) shows the PL spatial map of a sample of GNR **57** deposited on a coverslip. The AFM measurement of the corresponding area shows that this emission arises from small aggregates. These studies demonstrate that work on the solubilisation of GNRs is still needed in order to exploit and study their intrinsic optical properties.

Finally, despite the solution-mediated synthesis offering versatile techniques for the conception of GNRs, a limiting polymerization degree is generally reached due to a decrease of solubility over the polymerization process. Knowingly, surface assisted synthesis was developed with the aim of increasing the size of GNRs by bypassing the issue of solubility.

2.3 Surface-assisted synthesis of GNRs

2.3.1 Pure carbon GNR

According to the solution-mediated strategy, common synthetic routes of GNRs consisted in a first polymerization step of molecular precursors followed by a reaction of cyclodehydrogenation or graphenization to flatten the assembly and induced the aromatization of the system. In the same vein, the on-surface growth of GNRs required two sequential reaction steps: coupling reactions via a radical polymerization of the precursors followed by intramolecular cyclodehydrogenations to give the fully aromatic system. Historically, the feasibility of these thermally activated formation of nanostructures on surface was proved by the coupling between aryl halides porphyrin by Grill in 2007 [32] and the intramolecular cyclodehydrogenation was demonstrated by Fasel *et al* in 2010 on a triangular PAH [33].

In 2010, Fasel and Müllen developed a surface assisted protocol for the conception of armchair GNR (A-GNR) **58** and chevron-type GNR (c-GNR) **48** (figure 2.9) [34]. The protocol relies on the adsorption of the molecular precursors on a catalytic surface, generally on noble metals like Au, Ag, Cu, Pt, etc, followed by the radical polymerization triggered by thermal activation the surface at T_1 (around 200 °C), finally the cyclodehydrogenation occurred by annealing the surface at higher temperature T_2 (around 400 °C). The reaction was performed in a high vacuum chamber and the resulting was imaged by STM.

In this example, GNR **58** was synthesized by dehydrogenation of the polyanthryl precursor **60** which was formed on Au (111) by reaction of radicals derived from the 10,10'-dibromo-9,9'-bianthryl (DBBA) **61**. For the chevron-type GNR **48** (which can be also synthesized in solution—see section 2.2), the alternate polymer **50** was synthesized by reaction of the 6,11-dibromo-1,2,3,4-tetraphenyltriphenylene monomer **49**. Slightly higher temperatures were required for the synthesis of c-GNR ($T_1 = 250$ °C and $T_2 = 440$ °C instead of $T_1 = 200$ °C and $T_2 = 400$ °C for A-GNR). According to DFT calculation, a similar band gap of 1.6 eV was established for the two GNRs despite their different topologies. Later on, another investigation of the band gap using scanning tunneling spectroscopy (STS) revealed a band gap of 2.4-2.7 eV for A-GNR **58** on Au (111) [35, 36].

It was found that depending on the nature of the metal surface and its orientations, kinked GNR could be afforded. This observation was explained by a difference of reactivity and interaction between the molecules and the surface which impacted the diffusion of species and the stabilization of radicals during the polymerization process [37–41].

Moreover, by replacing bromines with chlorines, Jacobse *et al* [42] found that the intramolecular cyclodehydrogenation of 10,10'-dichloro-9,9'-bianthryl (DCBA) occurred at lower temperatures than the one required for the coupling-polymerization which gave rise to disordered polybisanthene as depicted in figure 2.10. In 2015, Olszowski *et al* [43] investigated the reactivity of 10,10'-diiodo-9,9'-bisanthryl (DIBA) on Ge (001) and found an activation temperature of the polymerization in the range of 200°C–250 °C, which was similar to the study on DBBA **61**. Other studies relating to the investigation of the effect of halogens can be found in the literature. The groups of Crommie and Fischer showed similar activation temperatures between the iodo or bromo version of the precursor **49** [44], while Grill *et al* succeeded in the selective thermal activation of iodines or bromines of a 5,15-bis(4-bromophenyl)-10,20-bis(4-iodophenyl)porphyrin using two distinct temperatures for triggering either linear or 2D structures on Au (111) [45].

Inspired by the work of Müllen and Fasel, many examples of original GNR structures grown on the surface have emerged during the last nine years. According to literature, armchair GNRs can be divided into three families: $N = 3p$ and $N = 3p + 1$ families (p is an integer and N the number of C–C lines in the width) with wide band gaps inversely proportional to the width of ribbons and the $N = 3p + 2$ family predicted to be nearly metallic with a very small band gap [46, 47]. Focusing on the variation of

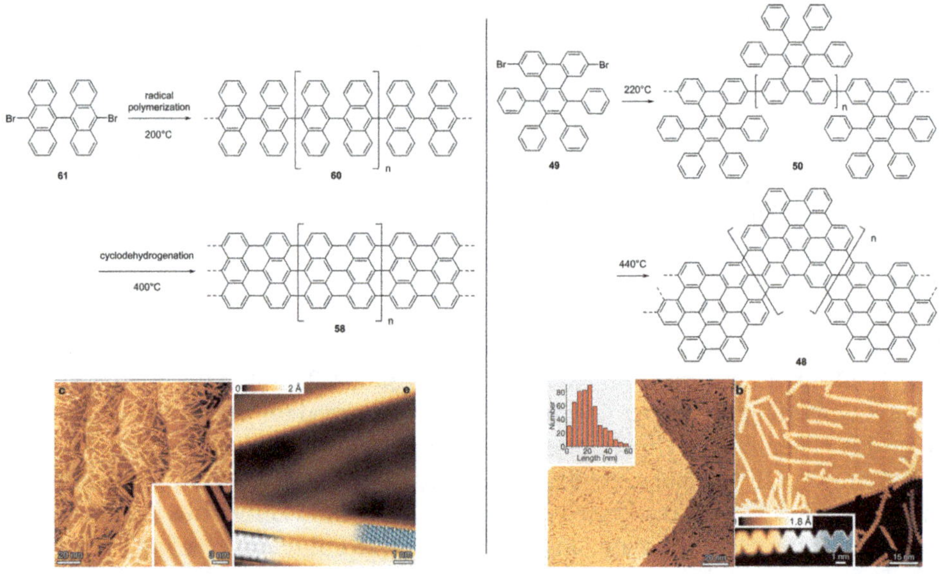

Figure 2.9. Surface-assisted protocol developed by Müllen and Fasel for the fabrication of straight and A-GNR 58 from DBBA precursor 61 and c-GNR 48 from precursor 49. STM images of GNRs 58 and 48 on Au (111) superimposed with calculated model in blue (bottom). Reprinted from [34], Copyright (2010), with permission from Springer Nature.

the width of A-GNRs, the group of Liljeroth reported the structure of the narrowest GNR **62** ($N = 5$) of the $N = 3p + 2$ family (figure 2.11(a)). GNR **62** was grown on Au (111) from dibromoperylene **63** and a very narrow band gap of 0.1 eV was determined by STS [48]. Concerning the $N = 3p$ and $N = 3p + 1$ families, in 2017, Fasel *et al* reported the structure of GNR **65** ($N = 9$) (figure 2.11(b)) which can be seen as a larger version of the A-GNR **58** ($N = 7$) and obtained from the dibromoterphenyl monomer **66** in Au (111) [49]. A band gap of 1.7 eV was determined by STS. Again, by re-designing DBBA monomer, the groups of Crommie and Fischer added two ranks of phenyls to the initial GNR **58** and formed a larger GNR **68** with a width of $N = 13$ on Au (111) (figure 2.11(c)) [50]. STS characterization exhibits a smaller gap around 1.4 eV. All of these experimental results showed the dependence between the width and the electronic properties of ribbons.

Although A-GNRs with different widths have been widely investigated, research towards the formation of alternative edges like zigzag- or gulf-types GNR suffer from low reporting. Zigzag GNRs (z-GNRs) are expected to possess magnetic properties thanks to their spin polarized edge states and might be promising for spintronic applications [51]. Intrinsically, ferromagnetic coupling is expected along the edges while antiferromagnetic comportment is expected between the opposite edges [52].

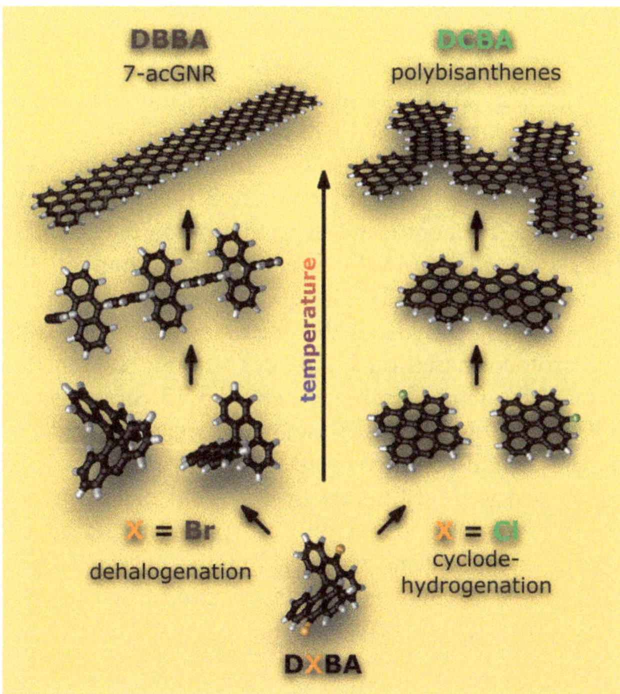

Figure 2.10. Representation of the different materials obtained on Au (111) for bromo- and chloro-substituted *bis*-anthracene precursor due to the inversion of activation temperature of dehalogenation and cyclodehydrogenation reactions. [42] courtesy of the authors.

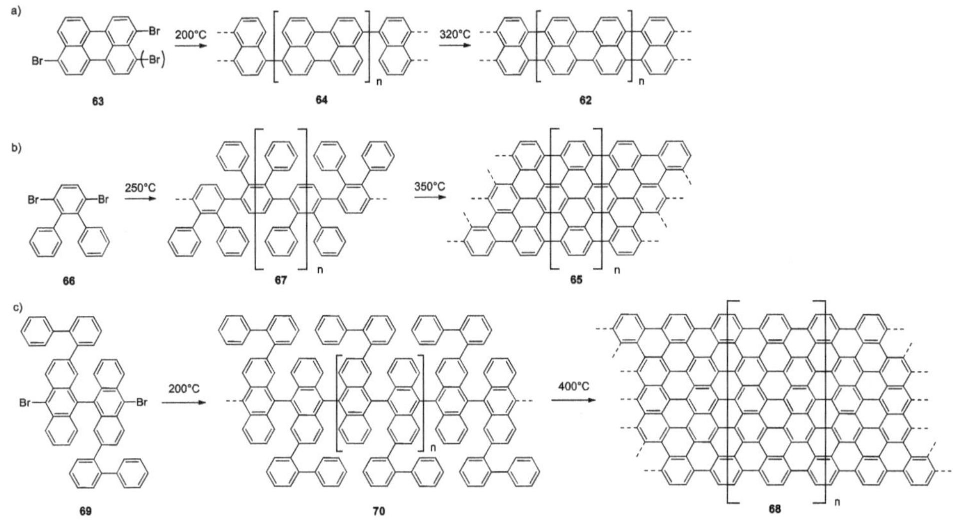

Figure 2.11. On surface synthesis on Au (111) of straight A-GNRs with different widths GNR 62 ($N = 5$) [48], GNR 65 ($N = 9$) [49] and GNR 68 ($N = 13$) [50].

The experimental visualization of these spin polarized edge states required highly uniform structure exhibiting atomic precision which cannot be achieved by the classical top-down methods. The formation of z-GNRs was first achieved on surface by Ruffieux *et al* [53] in 2016. Two GNRs **71a** and **71b** were synthesized by polymerization of particular monomers **72a** and **72b** exhibiting zigzag edges followed by deshydrogenation. Chemical structures, STM and non-contact AFM images of GNRs GNR **71a** and **71b** are displayed in figure 2.12. Despite their high structural quality, these z-GNRs did not confirm the prediction on magnetism yet.

Concerning gulf type edges GNRs, in 2015, the group of Müllen and Fasel investigated the structure of GNR **73** which was synthesized both in solution and on surface (figure 2.13) [54]. Starting from the precursor containing terbutyl groups, the solution mediated protocol based on Ullmann coupling reaction of the bischrysene derivative **74** produced di-, tetra-, hexa- and octamers **75**. Subsequent cyclodehydrogenation with DDQ/CF_3SO_3H on these isolated species yielded fused dimers and tetramers whereas hexa- and octamers were found partially fused. The band gaps of the dimers and tetramers were estimated by DFT calculation at 2.61 and 2.01 eV, respectively, and experimentally based on absorption studies at 2.36 and 1.90 eV, respectively. When the synthesis was performed on the surface, longer GNRs of up to 20 nm were formed on Au (111) by successive annealing at 160 °C and 360 °C (figure 2.13). A band gap of 1.70 eV was estimated for these ribbons by DFT calculation.

2.3.2 Heteroatom-containing GNR

Band gaps engineering of GNRs through the modifications of their width and edges or by doping with heteroelements is particularly attractive for applications in field effect transistors (FETs), for example. In the first part of this section, we described

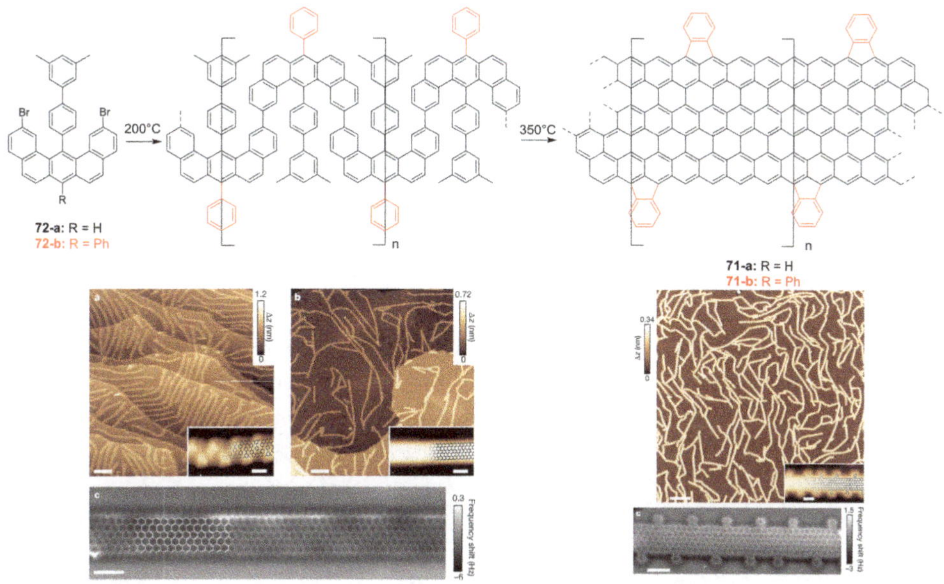

Figure 2.12. Synthetic route of z-GNRs 71a and 71b synthesized by on-surface protocol (top) and their corresponding STM and nc-AFM images on Au (111). Reprinted by permission from [53], © 2016 Springer Nature Publishing AG.

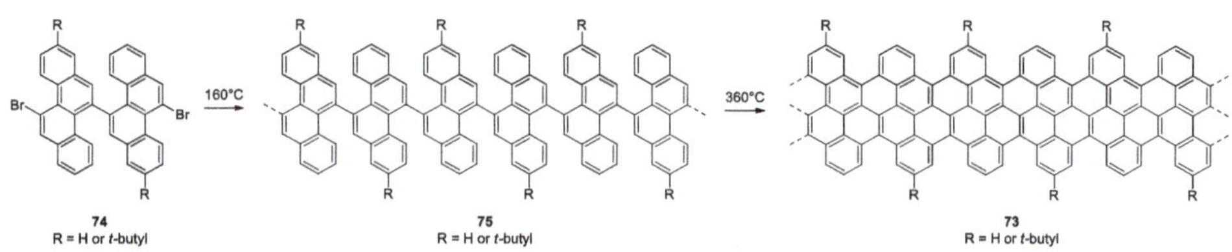

Figure 2.13. Synthetic route of gulf type GNR 73, synthesized in solution or on Au (111) by surface assisted protocol [54].

the different morphology of GNRs; in the following, we will focus on the fabrication nanoribbons containing heteroelements.

The effect of the concentration of the dopant on the band structures of GNRs has been investigated by Bronner *et al* [55] by comparison between a nitrogen-free chevron-type GNR **48** and its nitrogen-doped counterparts GNRs **75** and **76** (figures 2.14(a) and (b)). These c-GNRs were synthesized on Au (111) from different monomers: **49** for GNR **48** (see figure 2.9); **77** and **78** containing one or two pyridine rings for GNRs **75** and **76**, respectively. The characterization of the GNRs by ultraviolet photoelectron spectroscopy (UPS) and angle-resolved high-resolution electron energy loss spectroscopy (HREELS) demonstrated that although the band gaps were unaffected by the presence of nitrogen in the GNRs, the entire band structure was shifted to lower energy as a function of concentration of dopants. Such results indicate that band alignments can be achieved independently of the difference of energy between the valence and the conduction bands. Another example of

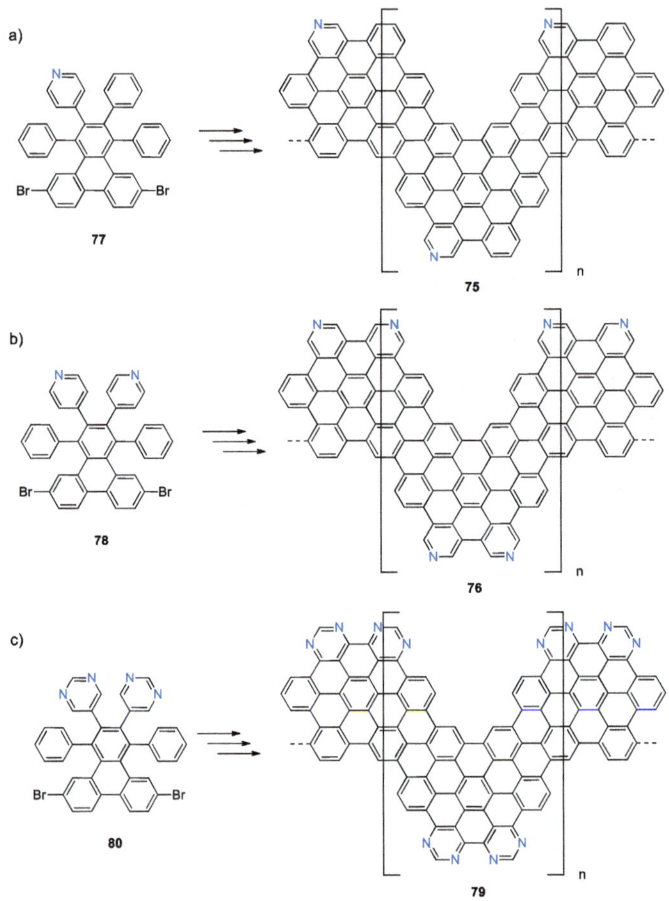

Figure 2.14. Structures of nitrogen-doped GNRs with variable nitrogen contents synthesized on Au (111). GNRs 75, 76 were reported by Bronner *et al* [55] and GNR 79 by Vo *et al* [56].

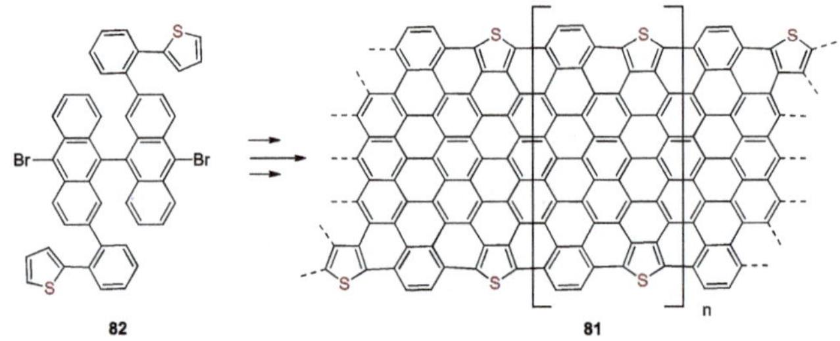

Figure 2.15. Surface synthesized sulphur-doped GNR **81** as reported by the groups of Crommie and Fischer. Because of the low difference of electronegativities between carbon and sulphur atoms, authors concluded that the effect of doping with sulphur did not strongly impact the position of the conduction band [57].

a nitrogen-doped structure with an increased nitrogen content was reported by Vo *et al* [56] (figure 2.14(c)). Indeed, c-GNR **79** was synthesized on Au (111) from monomer **80** containing two pyrimidine rings (four nitrogen atoms) per repeating units. According to STS studies, the comparison between GNR **48** and GNR **79** revealed similar band gaps close to 2 eV for both doped and undoped GNRs.

Another example of n-type doping was reported by the groups of Crommie and Fischer by incorporation of sulphur in the GNR. This sulphur-doped GNR **81** was synthesized on Au (111) from a precursor derived from DBBA containing two (2-phenyl)thiophene groups **82** (figure 2.15) [57]. As observed in the case of nitrogen doping, a similar band gap was estimated by comparison between the calculated band structures of sulphur-doped GNR **81** and its undoped version **68**. However, because of the small difference of electronegativity between carbon and sulphur atoms, the authors concluded that the effect of doping with sulphur did not strongly impact the position of the conduction band.

The p-doping of GNR with boron has been also developed simultaneously by Crommie and Fischer [58] and Kawai [59]. Both groups reported the synthesis of GNR **83** by on-surface polymerization of the boron-doped anthracenyl monomer **84**. The monomer was inspired by 10, 10'-dibromo-9, 9'-bianthryl (DBBA) in which a boronated anthracenoid moiety was introduced between the two bromoanthracenes (figure 2.16(a)). The electronic structure of the boron doped GNR **85** was investigated by both DFT calculation and STS analysis. The results based on DFT suggested that the doping introduced the acceptor band at energy within the gap compared to the undoped analogue GNR **58** (see figure 2.9). This new acceptor band became the novel conduction band, which significantly reduced the band gap of the boron-doped GNR from 2.1 to 0.8 eV [58]. On the other hand, STS measurement taking into account a contribution originating from the surface suggested that a stronger coupling occurred between the doped GNR and the substrate. Finally, a comparable band gap of 2.4 eV was estimated between doped and undoped GNR on Au (111) leaving the overall electronic structure unchanged [59]. It should be mentioned that Kawai *et al* further annealed the boron-doped GNRs and they were able to observe the fusion of their armchair GNR ($N = 7$) into larger armchair GNR ($N = 14$ and 21). In 2018, Wang *et al* [60] synthesized GNR **86** on Au (111) from precursor **87** containing with boronic bridges (figure 2.16(b)). The

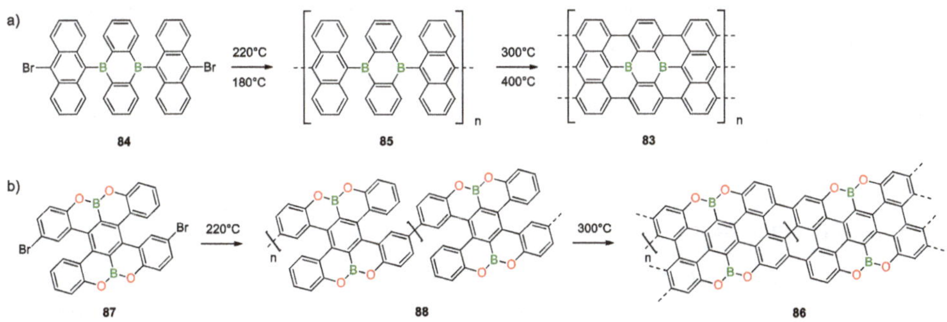

Figure 2.16. Boron-doped GNRs **83** [58, 59] and **86** [60] synthesized by surface assisted protocol.

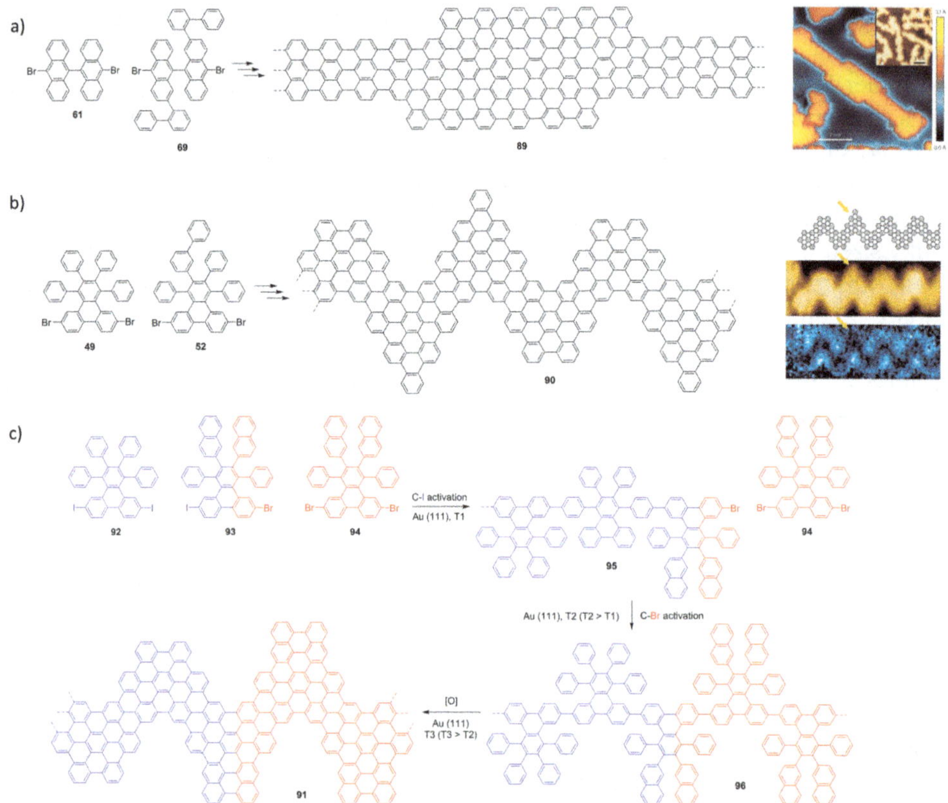

Figure 2.17. (a) Heterojunctions based on structural difference: A-GNR-segments (GNR 89) and corresponding STM images of the GNR junction [66], (b) c-GNR segments (GNR 90) with corresponding STM images of the GNR junction [67], (c) hierarchical method to fabricate heterojunction with controlled sequence (GNR 91) [68]. Reprinted by permission from Nature Nanotechnology [66], © 2015 Springer Nature Publishing AG. Reprinted from [67], Copyright (2018), with permission from Elsevier.

purpose of this work was to modulate the inter-ribbons interactions by influencing H bonds interactions. An experimentally band gap of *ca.* 3 eV was determined by STS and DFT.

From these studies, it turns out that several different factors like the morphology of the ribbon itself, the position of the dopants in the backbone and the doping ratio need to be considered to understand the effect of nitrogen or boron dopants on the electronic structure and measurements of doped ribbons in the FET configuration also need to be explored.

2.3.3 Heterojunctions

Finally, the idea to create thin and atomically precise heterojunctions in GNRs emerged for application in digital nanodevices, such as resonant tunnelling diodes or field effect transistors [61–63]. From the structural point of view, a heterojunction in

GNR is the fusion of two different GNR segments. These segments are generally made from precursors co-deposited on surface and exhibiting different widths or elemental composition [64, 65].

In 2015, the groups of Crommie and Fischer reported the synthesis of GNR **89** on Au (111) by co-deposition of molecular monomers **61** and **69** (figure 2.17(a)) [66]. These precursors were selected for their structural complementary based on the *bis*-anthracene backbone which allowed to form linear polymers and subsequently covalently bonded A-GNR segments with modulated widths of $N = 7$ and $N = 13$. According to STM observations, the widths of the narrower and of the wider segments were ca. 1.3 and 1.9 nm, respectively. Furthermore, according to tunnelling conductance measurement, the band gaps of each segment were estimated to be 2.76 ($N = 7$) and 1.57 eV ($N = 13$) which was in agreement with the previous reported values for corresponding GNR **58** ($N = 7$) and **68** ($N = 13$). According to experimental dI/dV maps along the junction correlated with DFT calculation, the authors suggested that the electronic structure of heterojunction **89** was similar to a type I semiconductor junction which implied that charge carriers can be trapped in the $N = 13$ segment because of the higher resistivity of the $N = 7$ segment.

Afterwards, Sinitskii reported an example of chevron type heterojunction in GNR **90** (figure 2.17(b)) [67]. The heterojunction was formed by co-deposition of monomers **49** and **52**; the presence of the additional phenyl in **52** allowed us to distinguish the constitutive segments of the GNR on STM images. Differential conductance dI/dV mapping at a voltage corresponding to the conduction band edge of c-GNRs (+1.6 eV) were performed on the heterojunction. It revealed high intensities on the edges of non-extended segments, while weaker signals were found in extremities of phenyl-extended segments. The presence of only one extra phenyl at the extremity of a segment, significantly affected its electronic structure and thus again demonstrated the necessity to atomically control the structure of these materials.

Still, the most critical aspect regarding the formation of heterojunctions remained the control of sequences and positions of heterojunctions. Indeed, the co-polymerization process does not permit control of the size of the different junctions and heterojunctions with random sequences and lengths are usually obtained.

In order to tackle these issues, the groups of Crommie and Fischer developed a hierarchical method to fabricate heterojunction with controlled sequence (GNR **91**—figure 2.17(c)) [68]. The strategy was based on the difference of thermal activation between C–I and C–Br bonds. Three monomeric precursors were synthesized: monomer **92** containing iodine atoms on each side, monomer **93** containing bromine and iodine atoms and monomer **94** (the dibromo version of the latest) and were sequentially deposited on Au(111) surface. In order to minimize the number of junction, monomer **93** (realizing the junction) was deposited in relatively low quantity compared to **92** and **94**. By heating the substrate at T_1, only polymerization of monomers containing iodine (**92** and **93**) occurred, which led to the formation of the first segment **95** terminated by extremities with bromine atoms. Then the substrate was annealed at a higher temperature T_2 to allow the brominated monomer **94** to react and give **96** which was then dehydrogenated to produce GNR **91**. The

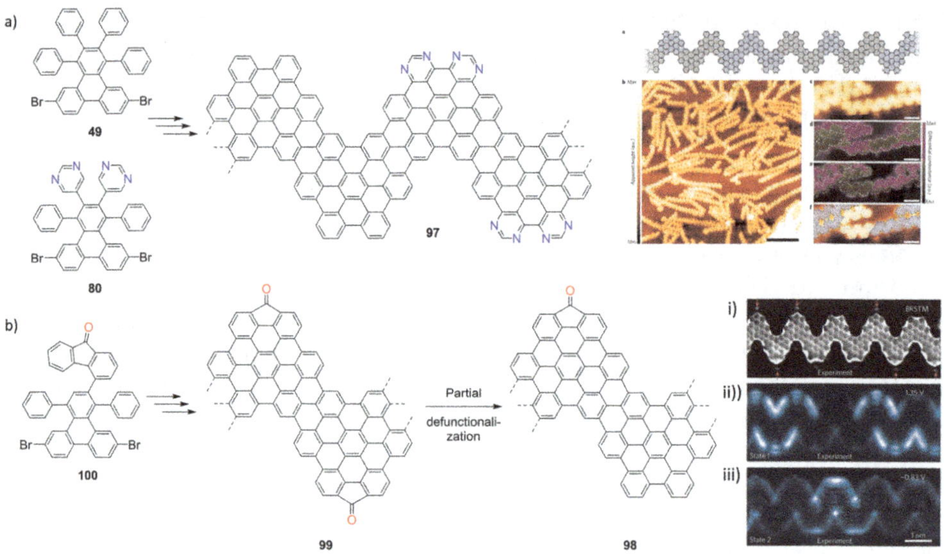

Figure 2.18. (a) Structures of chevron-type GNR 97 made of undoped and nitrogen-doped monomers and STM images and d*I*/d*V* maps with the nitrogen-doped segments highlighted in violet. (b) Structure of chevron-type GNR 98, the junction is obtained by selectively removing of carbonyl groups in the structure; on the right (i) bond resolved-STM image (a) differentiated via d*I*/d*V* map at energies corresponding to conduction (ii) and valence (iii) band edges. Reprinted by permission from Nature Nanotechnology [69], © 2014 Springer Nature Publishing AG. Reprinted by permission from Nature Nanotechnology [70], © 2017 Springer Nature Publishing AG.

heterojunction was proved by STM and STS: d*I*/d*V* mapping in addition to DFT calculation permitted to demonstrate the efficiency of the hierarchical strategy.

In 2014, the groups of Müllen and Fasel reported the fabrication of the GNR **97** containing heterojunctions by co-deposition and polymerization of monomers **49** and **80** (containing four nitrogens) (figure 2.18(a)) [69]. Because of the co-reaction, the resulting GNRs contained many junctions which were identified by differential conductance (d*I*/d*V*) maps. Finally, the group of Crommie and Fischer reported the formation of GNR **98** containing pure carbon and fluorenone parts on Au (111) (figure 2.18(b)) [70]. Their idea was to first synthesize a fully fluorenone functionalized GNR **99** by surface assisted polymerization and dehydrogenation of precursor **100** and then to partially remove carbonyl groups from the structure either by thermal treatment or by performing defunctionalization with the STM tip. The heterojunction was constituted by the combination of a fluorenone and a decarbonylated segments. According to DFT calculations and STS measurements, it was found that the band edges of fluorenone GNR segments were shifted to lower energy as compared with the one of non-functionalized segments and that the band alignment in GNR **98** yielded a type II heterojunction.

The optical properties of on-surface synthesized nanoribbons have been investigated by several groups. First, Denk *et al* [71] showed that the optical transitions of armchair GNR **58** (named 7-ANR) were dominated by excitonic effect. They performed reflectance difference spectroscopy experiments (RDS) on a film of

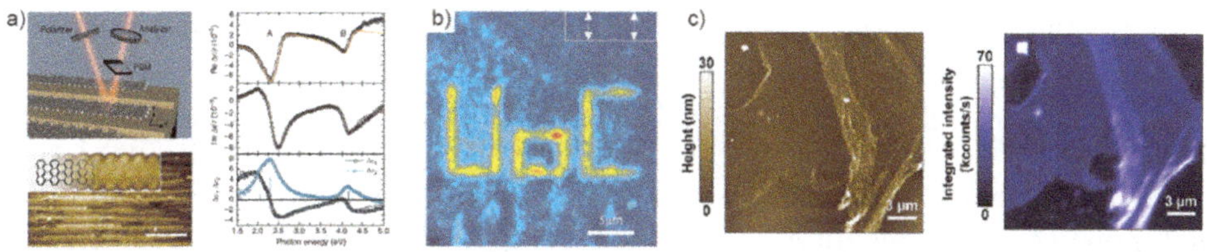

Figure 2.19. (a) Reflectance difference spectroscopy of 7-ANR 58 on Au [71]. (b) PL map of 7-ANR 58 on quartz after exposure to irradiation with a focused 440 nm laser [73]. (c) AFM and PL map of a 9-ANR 65 film transferred on a coverslip. Laser excitation at 405 nm [74]. (a) Reprinted by permission from Nature Communications [71], © 2014 Springer Nature Publishing AG. (b) Reprinted with permission from [73]. Copyright (2017) American Chemical Society. (c) [74] John Wiley & Sons. © 2017 WILEY-VCH Verlag GmbH & Co. KGaA, Weinheim.

oriented GNR **58** on gold (substrate used for the synthesis of the GNR) (figure 2.19(a)). The RDS spectra showed resonances at 2.1 and 2.3 eV that were attributed to excitons arising from optical transitions between the last valence and first conduction bands and the second last and second conduction bands. The authors also reported similar behaviour in chevron-type GNR **48** [72].

The investigation of the emission properties requires transferring the on-surface synthesized GNRs onto an isolating substrate. Indeed, the Au substrate used for the synthesis is supposed to quench the luminescence of the GNRs. Senkovskiy *et al* [73] investigated the photophysics of 7-armchair GNR **58** transferred on quartz. They found that the intrinsic photoluminescence of the GNR was low and that it could be enhanced by the creation of sp^3 defects by laser irradiation or hydrogenation in ultrahigh vacuum (figure 2.19(b)). Zhao *et al* [74] studied the photoluminescence properties of 7-armchair GNR **65** transferred on a glass coverslip. Figure 2.19(c) displays AFM and PL images of the exact same area of the GNR film. The direct comparison between the topography and the emission images shows that the PL is homogenously distributed on the GNR film. Here again, the PL was attributed to defect states created either during the synthesis or during the transfer process. Finally, Chong *et al* [75] have used another approach to study the luminescence of GNRs. They used a STM tip to lift a single 7-armchair GNR from the Au substrate. Then, they recorded the electroluminescence spectrum of the single GNR. They observed intrinsic luminescence arising from a transition between a Tamm state localized at the zig-zag terminus of the GNR and the top of the valence band.

2.4 Conclusion

In summary, we have presented here several examples of bottom-up synthesis of graphene nanoribbons. The synthesis can be divided in two categories: (i) the synthesis in solution and (ii) the on-surface synthesis. Both approaches exhibit advantages and drawbacks, for example, the synthesis in solution gives rise, in general, to longer GNR compared to on-surface synthesis and it permits us to have access to a relatively large amount of materials in solution, which are easier to manipulate. The main drawbacks of this method are the tendency of GNRs to aggregate in the solution during their synthesis and the potential presence of defects such as incomplete dehydrogenation or partial chlorination occurring during the Scholl reaction.

On-surface synthesis allows access to a larger variety of structures potentially less defective but in lower quantity. For many applications, these GNRs need to be transferred from the growth surfaces to insulating surfaces. This additional manipulation can constitute a limitation compared to the simple deposition of materials from the solution. Finally, it is worth mentioning that some GNR structures are accessible by the two methods (either in solution or on-surface).

Conversely to the top-down approach, the bottom synthesis of GNR permits to control at the atomic level the topology of the ribbons, the edge states and even the formation of heterojunctions. Despite the large attention drawn by the bottom-up synthesis of GNR, the characterization of their intrinsic properties is still limited. Finally, there is still a lot of work to do in order to clarify the intrinsic electronic and optical properties of GNRs, which is a prerequisite before one can take advantage of their promising assets for electronic, optoelectronic or spintronic devices.

References

[1] Jiao L, Zhang L, Wang X, Diankov G and Dai H 2009 Narrow graphene nanoribbons from carbon nanotubes *Nature* **458** 877–80

[2] Kosynkin D V, Higginbotham A L, Sinitskii A, Lomeda J R, Dimiev A, Price B K and Tour J M 2009 Longitudinal unzipping of carbon nanotubes to form graphene nanoribbons *Nature* **458** 872–76

[3] Novoselov K S, Fal'Ko V I, Colombo L, Gellert P R, Schwab M G and Kim K A 2012 Roadmap for graphene *Nature* **490** 192–200

[4] Stille J K, Noren G K and Green L 1970 Hydrocarbon ladder aromatics from a Diels–Alder reaction *J. Polym. Sci.* A **8** 2245–54

[5] Löffler M, Schlüter A D, Gessler K, Saenger W, Toussaint J M and Brédas J L 1994 Synthesis of a fully unsaturated 'molecular board' *Angew. Chemie Int. Ed. Engl.* **33** 2209–12

[6] Scherf U and Mullen K 1991 A soluble ladder polymer via bridging of functionalized poly(p-phenylene)-precursors *makromol Chem. Rapid Commun* **12** 489–97

[7] Scherf U and Müllen K 1992 Poly(arylenes) and poly(arylenevinylenes) 11 A modified two-step route to soluble phenylene-type ladder polymers *Macromolecules* **25** 3546–48

[8] Chmil K and Scherf U 1993 A simple two-step synthesis of a novel fully aromatic ladder-type polymer *Die Makromol. Chemie Rapid Commun* **14** 217–22

[9] Chmil K and Scherf U 1997 Conjugated all-carbon ladder polymers improved solubility and molecular weights *Acta Polym.* **48** 208–11

[10] Goldfinger M B and Swager T M 1994 Fused polycyclic aromatics via electrophile-induced cyclization reactions application to the synthesis of graphite ribbons *J. Am. Chem. Soc.* **116** 7895–96

[11] Shifrina Z B, Averina M S, Rusanov A L, Wagner M and Müllen K 2000 branched polyphenylenes by repetitive Diels-Alder cycloaddition *Macromolecules* **33** 3525–29

[12] Wu J, Gherghel L, Watson M D, Li J, Wang Z, Simpson C D, Kolb U and Müllen K 2003 From branched polyphenylenes to graphite ribbons *Macromolecules* **36** 7082–89

[13] Fogel Y, Zhi L, Rouhanipour A, Andrienko D, Räder H J and Müllen K 2009 Graphitic nanoribbons with dibenzo[e,l]pyrene repeat units synthesis and self-assembly *Macromolecules* **42** 6878–84

[14] Narita A *et al* 2014 Synthesis of structurally well-defined and liquid-phase-processable graphene nanoribbons *Nat. Chem.* **6** 126–32

[15] Narita A *et al* 2014 Bottom-up synthesis of liquid-phase-processable graphene nanoribbons with near-infrared absorption *ACS Nano* **8** 11622–630

[16] Yang X, Dou X, Rouhanipour A, Zhi L, Räder H J and Müllen K 2008 Two-dimensional graphene nanoribbons *J. Am. Chem. Soc.* **130** 4216–17

[17] Li G, Yoon K Y, Zhong X, Zhu X and Dong G 2016 Efficient bottom-up preparation of graphene nanoribbons by mild suzuki–miyaura polymerization of simple triaryl monomers *Chem. - A Eur. J.* **22** 9116–120

[18] Dössel L, Gherghel L, Feng X and Müllen K 2011 Graphene nanoribbons by chemists nanometer-sized soluble and defect-free *Angew. Chemie. Int. Ed.* **50** 2540–543

[19] Yang W, Lucotti A, Tommasini M and Chalifoux W A 2016 Bottom-up synthesis of soluble and narrow graphene nanoribbons using alkyne benzannulations *J. Am. Chem. Soc.* **138** 9137–144

[20] Gao J, Uribe-Romo F J, Saathoff J D, Arslan H, Crick C R, Hein S J, Itin B, Clancy P, Dichtel W R and Loo Y L 2016 Ambipolar transport in solution-synthesized graphene nanoribbons *ACS Nano* **10** 4847–856

[21] Odian G 2004 *Principles of Polymerization* 4th edn (New York: Wiley)

[22] Schmidt J, Werner M and Thomas A 2009 Conjugated microporous polymer networks via Yamamoto polymerization *Macromolecules* **42** 4426–29

[23] Schwab M G, Narita A, Hernandez Y, Balandina T, Mali K S, De Feyter S, Feng X and Müllen K 2012 Structurally defined graphene nanoribbons with high lateral extension *J. Am. Chem. Soc.* **134** 18169–72

[24] Vo T H,, Shekhirev M, Kunkel D A, Morton M D, Berglund E, Kong L, Wilson P M, Dowben P A, Enders A and Sinitskii A 2014 Large-scale solution synthesis of narrow graphene nanoribbons *Nat. Commun.* **5** 3189

[25] Pour M M *et al* 2017 Laterally extended atomically precise graphene nanoribbons with improved electrical conductivity for efficient gas sensing *Nat. Commun.* **8** 1–9

[26] Vo T H, Shekhirev M, Kunkel D A, Orange F, Guinel M J F, Enders A and Sinitskii A 2014 Bottom-up solution synthesis of narrow nitrogen-doped graphene nanoribbons *Chem. Commun.* **50** 4172–74

[27] Huang Y *et al* 2016 Poly(ethylene oxide) functionalized graphene nanoribbons with excellent solution processability *J. Am. Chem. Soc.* **138** 10136–39

[28] Keerthi A *et al* 2017 Edge functionalization of structurally defined graphene nanoribbons for modulating the self-assembled tructures *J. Am. Chem. Soc.* **139** 16454–57

[29] Slota M *et al* 2018 Magnetic edge states and coherent manipulation of graphene nanoribbons *Nature* **557** 691–95

[30] Tan Y Z, Yang B, Parvez K, Narita A, Osella S, Beljonne D, Feng X and Müllen K 2013 Atomically precise edge chlorination of nanographenes and its application in graphene nanoribbons *Nat. Commun.* **4** 2646

[31] Zhao S *et al* 2017 Fluorescence from graphene nanoribbons of well-defined structure *Carbon* **119** 235–40

[32] Grill L, Dyer M, Lafferentz L, Persson M, Peters M V and Hecht S 2007 Nano-architectures by covalent assembly of molecular building blocks *Nat. Nanotechnol.* **2** 687–91

[33] Treier M, Pignedoli C A, Laino T, Rieger R, Müllen K, Passerone D and Fasel R 2011 Surface-assisted cyclodehydrogenation provides a synthetic route towards easily processable and chemically tailored nanographenes *Nat. Chem.* **3** 61–7

[34] Cai J *et al* 2010 Atomically precise bottom-up fabrication of graphene nanoribbons *Nature* **466** 470–73

[35] Ruffieux P *et al* 2012 Electronic structure of atomically precise graphene nanoribbons *ACS Nano* **6** 6930–35

[36] Koch M, Ample F, Joachim C and Grill L 2012 Voltage-dependent conductance of a single graphene nanoribbon *Nat. Nanotechnol.* **7** 713–17

[37] Schulz F *et al* 2017 Precursor geometry determines the growth mechanism in graphene nanoribbons *J. Phys. Chem.* C **121** 2896–904

[38] Han P, Akagi K, Federici Canova F, Mutoh H, Shiraki S, Iwaya K, Weiss P S, Asao N and Hitosugi T 2014 Bottom-up graphene-nanoribbon fabrication reveals chiral edges and enantioselectivity *ACS Nano* **8** 9181–87

[39] De Oteyza D G *et al* 2016 Substrate-independent growth of atomically precise chiral graphene nanoribbons *ACS Nano* **10** 9000–08

[40] Simonov K A, Generalov A V, Vinogradov A S, Svirskiy G I, Cafolla A A, McGuinness C, Taketsugu T, Lyalin A, Mårtensson N and Preobrajenski A B 2018 Synthesis of armchair graphene nanoribbons from the 10,10′-dibromo-9,9′-bianthracene molecules on Ag(111): the role of organometallic intermediates *Sci. Rep.* **8** 3506

[41] Ohtomo M, Jippo H, Hayashi H, Yamaguchi J, Ohfuchi M, Yamada H and Sato S 2018 Interpolymer self-assembly of bottom-up graphene nanoribbons fabricated from fluorinated precursors *ACS Appl. Mater. Interfaces* **10** 31623–30

[42] Jacobse P H, van den Hoogenband A, Moret M E, Klein Gebbink R J M and Swart I 2016 Aryl radical geometry determines nanographene formation on Au(111) *Angew Chemie. Int. Ed.* **55** 13052–55

[43] Olszowski P, Zapotoczny B, Pe D and Guitia E 2015 Aryl halide C–C coupling on Ge(001): H surfaces *J. Phys. Chem.* C **119** 27478–82

[44] Bronner C, Marangoni T, Rizzo D J, Durr R A, Jørgensen J H, Fischer F R and Crommie M F 2017 Iodine versus bromine functionalization for bottom-up graphene nanoribbon growth: role of diffusion *J. Phys. Chem.* C **121** 18490–95

[45] Lafferentz L, Eberhardt V, Dri C, Africh C, Comelli G, Esch F, Hecht S and Grill L 2012 Controlling on-surface polymerization by hierarchical and substrate-directed growth *Nat. Chem.* **4** 215–20

[46] Brey L and Fertig H A 2006 Electronic states of graphene nanoribbons studied with the Dirac equation *Phys. Rev.* B **73** 2–6

[47] Yang L, Park C H, Son Y W, Cohen M L and Louie S G 2007 Quasiparticle energies and band gaps in graphene nanoribbons *Phys. Rev. Lett.* **99** 186801

[48] Kimouche A, Ervasti M M, Drost R, Halonen S, Harju A, Joensuu P M, Sainio J and Liljeroth P 2015 Ultra-narrow metallic armchair graphene nanoribbons *Nat. Commun.* **6** 10177

[49] Talirz L *et al* 2017 On-surface synthesis and characterization of 9-atom wide armchair graphene nanoribbons *ACS Nano* **11** 1380–88

[50] Chen Y C, De Oteyza D G, Pedramrazi Z, Chen C, Fischer F R and Crommie M F 2013 Tuning the band gap of graphene nanoribbons synthesized from molecular precursors *ACS Nano* **7** 6123–28

[51] Cummings A W, Valenzuela S O, Ortmann F and Roche S 2017 Graphene spintronics *2D Materials Properties and Devices* eds P Avouris, T F Heinz and T Low (Cambridge: Cambridge University Press), pp 197–218
[52] Nakada K, Fujita M, Dresselhaus G and Dresselhaus M S 1996 Edge state in graphene ribbons: nanometer size effect and edge shape dependence *Phys. Rev.* B **54** 17954–961
[53] Ruffieux P *et al* 2016 On-surface synthesis of graphene nanoribbons with zigzag edge topology *Nature* **531** 489–92
[54] Liu J, Li B-W, Tan Y-Z, Giannakopoulos A, Sanchez-Sanchez C, Beljonne D, Ruffieux P, Fasel R, Feng X and Müllen K 2015 Toward cove-edged low band gap graphene nanoribbons *J. Am. Chem. Soc.* **137** 6097–103
[55] Bronner C, Stremlau S, Gille M, Brauße F, Haase A, Hecht S and Tegeder P 2013 Aligning the band gap of graphene nanoribbons by monomer doping *Angew. Chemie. Int. Ed.* **52** 4422–25
[56] Vo T H *et al* 2015 Nitrogen-doping induced self-assembly of graphene nanoribbon-based two-dimensional and three-dimensional metamaterials *Nano Lett.* **15** 5770–77
[57] Nguyen G D *et al* 2016 Bottom-up synthesis of $N = 13$ sulfur-doped graphene nanoribbons *J. Phys. Chem.* C **120** 2684–87
[58] Cloke R R, Marangoni T, Nguyen G D, Joshi T, Rizzo D J, Bronner C, Cao T, Louie S G, Crommie M F and Fischer F R 2015 Site-specific substitutional boron doping of semiconducting armchair graphene nanoribbons *J. Am. Chem. Soc.* **137** 8872–75
[59] Kawai S, Saito S, Osumi S, Yamaguchi S, Foster A S, Spijker P and Meyer E 2015 Atomically controlled substitutional boron-doping of graphene nanoribbons *Nat. Commun.* **6** 8098
[60] Wang X Y *et al* 2018 Bottom-up synthesis of heteroatom-doped chiral graphene nanoribbons *J. Am. Chem. Soc.* **140** 9104–07
[61] Llinas J P *et al* 2017 Short-channel field-effect transistors with 9-atom and 13-atom wide graphene nanoribbons *Nat. Commun.* **8** 633
[62] Yoon Y and Salahuddin S 2012 Dissipative transport in rough edge graphene nanoribbon tunnel transistors *Appl. Phys. Lett.* **101** 263501
[63] Yoon Y and Salahuddin S 2010 Barrier-free tunneling in a carbon heterojunction transistor *Appl. Phys. Lett.* **97** 0331102
[64] Rizzo D J, Veber G, Cao T, Bronner C, Chen T, Zhao F, Rodriguez H, Louie S G, Crommie M F and Fischer F R 2018 Topological band engineering of graphene nanoribbons *Nature* **560** 204–8
[65] Gröning O *et al* 2018 Engineering of robust topological quantum phases in graphene nanoribbons *Nature* **560** 209–13
[66] Chen Y C, Cao T, Chen C, Pedramrazi Z, Haberer D, De Oteyza D G, Fischer F R, Louie S G and Crommie M F 2015 Molecular bandgap engineering of bottom-up synthesized graphene nanoribbon heterojunctions *Nat. Nanotechnol.* **10** 156–60
[67] Costa P S, Teeter J D, Enders A and Sinitskii A 2018 Chevron-based graphene nanoribbon heterojunctions: localized effects of lateral extension and structural defects on electronic properties *Carbon* **134** 310–15
[68] Bronner C *et al* 2018 Hierarchical on-surface synthesis of graphene nanoribbon heterojunctions *ACS Nano* **12** 2193–200
[69] Cai J *et al* 2014 Graphene nanoribbon heterojunctions *Nat. Nanotechnol.* **9** 896–900

[70] Nguyen G D *et al* 2017 Atomically precise graphene nanoribbon heterojunctions from a single molecular precursor *Nat. Nanotechnol.* **12** 1077–82
[71] Denk R *et al* 2014 Exciton-dominated optical response of ultra-narrow graphene nanoribbons *Nat. Commun.* **5** 5253
[72] Denk R *et al* 2017 Probing optical excitations in chevron-like armchair graphene nanoribbons *Nanoscale* **9** 18326–33
[73] Senkovskiy B V, Pfeiffer M, Alavi S K, Bliesener A, Zhu J, Michel S, Fedorov A V, German R, Hertel D, Haberer D, Petaccia L, Fischer F R, Meerholz K, van Loosdrecht P H M, Lindfors K and Grüneis A 2017 Making graphene nanoribbons photoluminescent *Nano Lett.* **17** 4029–37
[74] Zhao S *et al* 2017 Optical investigation of on-surface synthesized armchair graphene nanoribbons *Phys. Status Solidi Basic Res.* **254** 1700223
[75] Chong M C, Afshar-imani N, Scheurer F, Cardoso C, Ferretti A, Prezzi D and Schull G 2018 Bright electroluminescence from single graphene nanoribbon junctions *Nano Lett.* **18** 175–81

IOP Publishing

Graphene Nanoribbons

Luis Brey, Pierre Seneor and Antonio Tejeda

Chapter 3

Spin–orbit in graphene nanoribbons

M Pilar López Sancho and M Carmen Muñoz

Kane and Mele predicted that the intrinsic spin–orbit coupling leads to realizing the quantum spin Hall insulator phase in graphene. As a consequence, spin-filtered edge states would carry dissipationless spin current in graphene nanoribbons. Here, the effect of the intrinsic spin–orbit coupling in graphene nanoribbons with different edge shapes and widths are investigated within the four-orbital tight-binding model. Symmetry is found to be determinant in the induced spin–orbit coupling effects. In graphene nanoribbons supporting edge states, the spin–orbit interaction (SOI) gives rise to spin-filtered 1D-helical states fully localized at the edges.

3.1 Introduction

At the beginning of the graphene rise, Kane and Mele predicted that the intrinsic spin–SOI drives a topological phase in graphene [1, 2]. Although the topological phase has not been observed in graphene, in the last years, it has been undoubtedly shown that the SOI is an essential ingredient to realize time-reversal invariant topological insulators [3]. Graphene is a two-dimensional (2D) sheet of carbon atoms characterized by its massless low-energy excitations. It has outstanding electrical and mechanical properties [4] exhibiting numerous intriguing topological phenomena, such as the integer and the fractional quantum Hall effects [5, 6]. Carbon nanotubes (CNTs), formed by rolling a graphene sheet onto a cylinder, and graphene stripes, known as graphene nanoribbons (GNRs), are quasi-1D systems with tunable electronic properties. They hold great promise for nanoelectronics and other technologies, such as spintronic or quantum computing [7, 8]. Chirality and diameter determine the electronic properties of CNTs, while the edge shape and ribbon width define those of GNR. In particular, the edge shape of the ribbon leads to strikingly different properties of the states near the Fermi level. Ribbons with zigzag edges possess partly flat bands at the Fermi level corresponding to electronic states localized in the near vicinity of the edge. In contrast, localized edge states and

doi:10.1088/978-0-7503-1701-6ch3 © IOP Publishing Ltd 2020

the corresponding flat bands are completely absent for ribbons with armchair edges, which show a metallic or semiconductor behaviour depending on their width.

Experimental investigations of GNR had been hampered because the nanoribbon edges could not be produced with atomic precision and the graphene terminations that had been proposed were chemically unstable [9, 10]. However, recent advancements in the bottom-up molecular synthesis of GNRs have allowed for the fabrication of GNRs with atomic precision control of their edge and width, and with various structures uniquely determined by the molecular precursor [11–13]. The atomically precise GNRs synthesized using the bottom-up precursor molecule techniques are of nanometre-scale widths and present a variety of termination patterns, from armchair to cove or chevron-edged and hence different electronic state. Their growth has allowed the exploration of GNR electronic properties, such as the energy-gap versus width relations [14]. Moreover, recent theoretical work has predicted the existence of 1D symmetry-protected topological phases in semiconductor graphene nanoribbons [15]. The topological phase of these laterally confined semiconducting strips of graphene is determined by their edge shape, width and terminating crystallographic unit cell and is characterized by a Z_2 invariant [16]. The topological phases have been experimentally investigated in semiconducting armchair AGNRs [17–20].

Because of the Dirac nature of its charge carriers and the presence of two valleys, spin–orbit induced topological order and the quantum spin Hall effect were first proposed in graphene. Symmetry allowed SOIs can generate an energy gap and convert graphene from a two-dimensional zero gap semiconductor to an insulator with a quantized spin Hall effect. The intrinsic spin–orbit coupling is responsible for the non-trivial topological order [1, 2]. Nevertheless, to this date, topological states have not been observed experimentally in graphene probably due to the weak strength of the graphene intrinsic SOI. However, SOI effects had been previously investigated in CNTs. The curvature of the surface breaks certain symmetries present in the planar geometry and this broken symmetry enhances the intrinsic SOI in carbon nanotubes compared with flat graphene. It had been shown theoretically that in metallic CNTs an energy gap opens for the energy bands crossing at the Fermi energy, while in semiconductor CNTs the energy bands split by the SOI. The built-in effective magnetic field is aligned along the nanotube axis [21–23]. In addition, the chirality dependence of the SOI, was predicted. SOI lifts the spin degeneracy of energy bands for chiral CNT, albeit the spin degeneracy survives for achiral tubes. The absence of an inversion center in the chiral CNT being responsible for the spin-splitting [22]. Later transport measurements in CNTs reported asymmetric splitting between conduction and valence bands as well as diameter and chirality dependence of the gap and energy splitting. The SOI splittings are of the order of meVs -a zero-field spin splitting of up to 3.4 meV was found in CNT devices [24–28]. These values are orders of magnitude larger than that predicted in graphene. Although most of the characteristics of the measured SOI effects in CNT, were explained by four-orbital tight-binding simulations [29–31], the origin of the large SOI value remains unclear.

Due to the low atomic number of carbon the intrinsic SOI in graphene flat structure is weak and the intrinsic SO gap value has been elusive. This year [32], a zero-field splitting has been observed in Hall bar graphene structures, at low temperature, interpreted as an intrinsic SOI gap of 42.2 μeV. However, several approaches for enhancing the SOI in graphene have been proposed in the last years, such as external electric fields, adsorbed atoms on the graphene surface, the presence of strain or the proximity effect to transition metals. Recently, both, by endowing graphene with certain heavy adatoms and by exploiting the interfacial interactions between graphene and a semiconducting transition metal dichalcogenide, WS_2, a drastic increase of the SOI strength up to a few meV, without altering the graphene electronic structure, has been advanced [33–35]. Accordingly, the actual capability to grow GNRs with controlled edge-shape and width at the atomic level together with the possibility to induce an enhancement of the intrinsic SOI, bring the prospect to realize new topological insulating states and provide a mechanism for all-electrical control of spins in graphene-based nanostructures [33]. Here, we analyze the effect of the intrinsic SOIs in graphene nanoribbons of different edge-shapes and widths with the final aim to unambiguously establish the nature of the SOI induced states for significant and plausible values of the effective intrinsic SOI in graphene, reachable through adatom layers or proximity effects.

3.2 Model and method

Graphene nanoribbons are 1D nanometer-wide strips of graphene with translational invariance along the longitudinal edge direction. The shape of the edges has important consequences for the electronic properties of the GNRs. Although generic ribbon borders are combination of different edges, there are two prototypical terminations of the GNRs, zigzag and armchair, with a $\theta = 30°$ angle of difference in the cutting direction of the two-dimensional graphene sheet. The width of the GNR is defined by n, the number of zigzag lines or the number of dimer lines for the zigzag (ZGNR) and armchair (AGNR) ribbons, respectively. The corresponding unit cell contains $N = 2n$ carbon atoms. Figure 3.1 shows schematic examples of the two ribbon types, both possess an inversion center. No reconstruction or relaxation of the edges are considered, since experimental results support the stability of the unrelaxed edges.

The electronic properties of the graphene nanoribbons are calculated from the one-electron Hamiltonian given by,

$$H = H_0 + H_{SO}$$

H_0 is the spin-independent non-interacting Slater-Koster empirical tight-binding (SK-ETB) Hamiltonian, and H_{SO} is the microscopic SOI term. A four-orbital, $2s$, $2p_x$, $2p_y$ and $2p_z$, basis set is considered in order to include the conventional on-site approach for the intrinsic SO interaction. Within this approximation the H_0 term is written as,

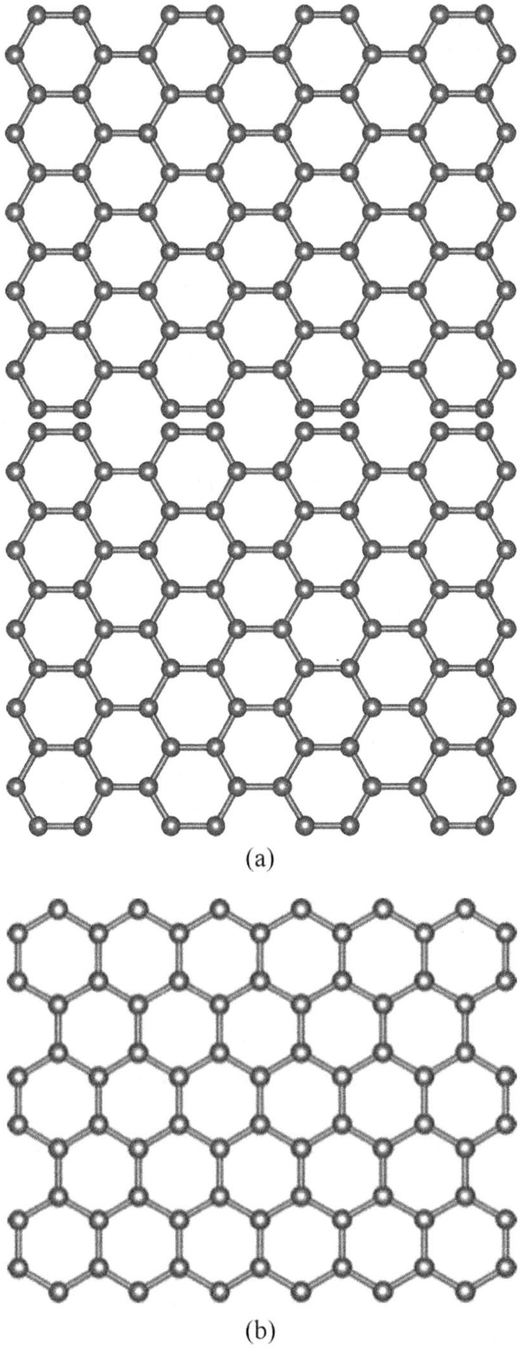

Figure 3.1. Schematic lattice structure of graphene nanoribbons with: (a) zigzag, and (b) armchair edges.

$$H_0 = \sum_{i,\alpha,s} \varepsilon_{i,s}^{\alpha} + \sum_{<ij>,\alpha,\beta,s} t_{ij}^{\alpha,\beta} c_{i,s}^{\alpha+} c_{j,s}^{\beta} + H.c., \tag{3.1}$$

where ϵ^{α} represents the atomic energy of the orbital α, $<ij>$ stands for atomic sites of the honeycomb lattice and $c_{i,s}^{\alpha+}$ and $c_{i,s}^{\alpha}$ are the creation and annihilation operators of one electron at site i, orbital α and spin s, respectively.

Since the major contribution of the crystal potential, ∇V, is near the atomic nuclei, the SOI can be accurately approximated by a local atomic term of the form:

$$H_{SO} = \sum_i \frac{\hbar}{4m^2c^2} \frac{1}{r_i} \frac{dV_i}{dr_i} \mathbf{L} \cdot \mathbf{S} = \lambda \mathbf{L} \cdot \mathbf{S},$$

where spherical symmetry of the atomic potential is assumed. r_i is the radial coordinate with origin at the atom i, V_i the spherical symmetric potential about the same atom and **L** the orbital angular momentum. The atomic SO coupling constant λ depends on the orbital angular momentum **L** and thus on the atomic orbital.

For the sp^3 model Hamiltonian, SO coupling occurs only among p orbitals and neglecting nearest neighbour SO terms, the intrinsic SOI is described as an on-site interaction among the p-orbitals. Thus, the H_{SO} term couples p orbitals on the same atom. The on-site approach is reasonable, due to the small atomic number of carbon atoms. In contrast, in the $k.p$ and effective π-band models, the atomic SOI is treated by perturbation theory as a second-neighbour spin dependent hopping term [21, 36].

H_{SO} adds diagonal and off-diagonal spin-dependent matrix elements to the $8N \times 8N$ matrix Hamiltonian, where N is the number of carbon atoms of the unit cell and 8 comes from the four-orbital per spin tight-binding (TB) basis set. Using the raising and lowering angular momentum operators, $L_+ = L_x + iL_y$ and $L_- = L_x - iL_y$, respectively and the Pauli spin matrices, the complete Hamiltonian, H, in the 2×2 block spinor structure is given by,

$$H = \begin{pmatrix} H_0 + \lambda L_z & \lambda L_- \\ \lambda L_+ & H_0 - \lambda L_z \end{pmatrix}. \tag{3.2}$$

where the atomic-like spin–orbit term H_{SO} has been added to SK-ETB spin-independent H_0 Hamiltonian. The diagonal terms act as an effective Zeeman field producing gaps of opposite signs at the K and K' points of the BZ. In the sp^3 TB model the non-vanishing matrix elements of H_{SO} couple $2p_x$, $2p_y$ and $2p_z$ orbitals. These terms have been widely discussed previously, see for example [37, 38]. H_{SO} induces $\sigma - \pi$ hybridization, does not break the equivalence of the two carbon sublattices and respect parity and time reversal symmetry, therefore spin degeneracy can not be removed on GNRs with an inversion center.

The electronic properties of the GNRS are obtained by the exact diagonalization of the total Hamiltonian H. The Tománek-Louie parameterization for graphite [39], previously used to calculate SOI effects on CNTs, has been also employed in the present calculations. The TB parameters, taken from [39], are: $\epsilon_s = -7.3$ eV, $\epsilon_p = 0.0$ eV, $ss\sigma = -4.30$ eV, $sp\sigma = 4.98$ eV, $pp\sigma = 6.38$ eV, $pp\pi = -2.66$ eV, and specify the energy scale of the model. The value of the spin–orbit coupling strength of graphene is unknown and some controversy exists with regards to its precise magnitude. The original estimate of 100 μV, was subsequently reduced to $\approx 1\ \mu V$, while first principles calculations reports values of 25–50 μV [1, 36, 40, 41]. Recent experiments have found evidence for enhanced effective SOI on the order of $\approx 1 - 15$ meV, but its precise nature remains uncertain [33, 42–44]. For the sake of clarity of the figures, we present here results obtained with a value of λ of 0.4 eV, which corresponds to ≈ 0.2, in units of the $pp\pi = -2.66$ eV parameter.

3.3 Results

Geometry changes essentially the physical properties of GNRs. Due to the finite size and the presence of boundaries, GNRs present different band structures depending on the atomic termination and on the width of the ribbon. The truncation of inter-atomic bonds caused by the borders gives rise to the appearance of edge states only for specific terminations and the energy subbands associated with the intrinsic band structure of the graphene sheet are also dependent on the boundary conditions of the GNR. Here we present results for GNRs of the two main geometries, ZGNRs and AGNRs, and different widths.

3.3.1 Zigzag graphene nanoribbons

In the zigzag termination, all atoms of each boundary belong to the same sublattice, and opposite edge atoms belong to different sublattice. Figure 3.2 presents the band structure of two ZGRNs with different widths, $N = 60$ and $N = 20$ with $\lambda = 0.4$ eV. In the absence of SOI the most remarkable feature of the band structure of a ZGNR—not shown in figure 3.2—is the emergence of the well known dispersionless π-derived states at zero energy [45–49]. Their wave functions are localized at the borders of the ribbon, form flat bands for $\frac{2}{3}\frac{\pi}{a} < k < \frac{\pi}{a}$, a being the graphene lattice constant. and give rise to a peak in the density of states at the Fermi energy being crucial for magnetic instabilities [47]. Hereafter $\mathbf{k}$ will be given in units of $\frac{\pi}{a}$. Besides, expanding the complete one-dimensional Brillouin zone (BZ) there is another edge state derived from the σ-orbitals, missed in one-band calculations, which has larger dispersion and overlaps with the subbands originating from the graphene sheet. Nevertheless, it occupies a k-gap in a large k interval of the BZ and thus its wavefunction is also fully localized at the edge atoms for k values within the ZGNR gaps. Gapless π-edge states are located at zero energy, as has been repeatedly predicted in previous $k.p$ and effective π-band model calculations [45–49], and the σ-edge band emerges around 1 eV at $|\mathbf{k}| = 0$. It is worth mentioning that edge states appear at equivalent energy independent of the ribbon width. For ZGNRs the valley character is preserved and both, π and σ-edge states, have a fourfold degeneracy associated with the spin and the existence of two identical edges. Even for narrow ZGRN there is not inter-edge coupling on account of its finite transverse size and the degeneracy associated with the two edges remains. In real ribbons the atoms of the edges are usually passivated, therefore the energy as well as the degeneracy of the σ-edge states should be different from those of the unsaturated bonds.

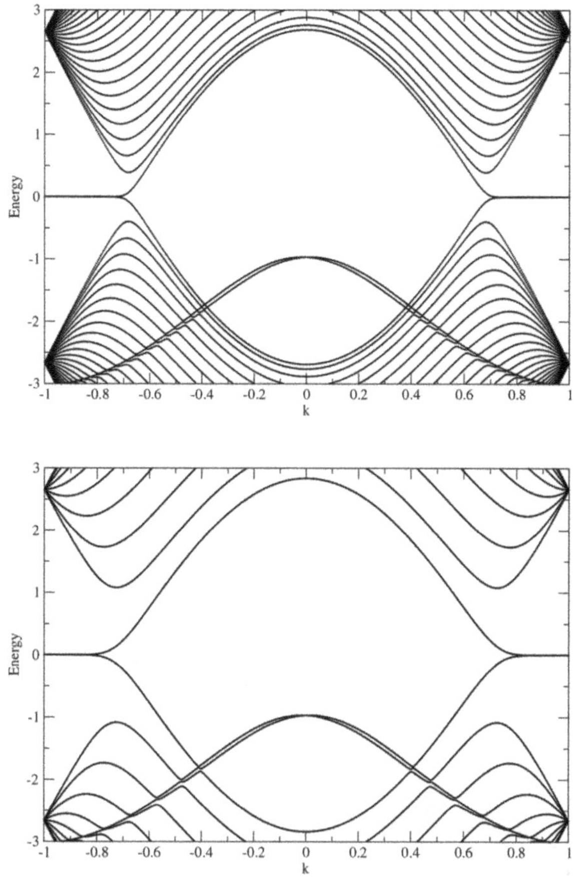

Figure 3.2. Band structure of the zigzag ribbon $N = 60$ (top) and $N = 20$ (bottom) with SOI $\lambda = 0.4$ eV. k are in units of $\frac{\pi}{a}$ and energies in eV.

When the SOI term is included the major effect is on the edge states. In both ZGNRs all the bands become twofold degenerate -because flat ribbons have an inversion center-, the spectrum remains gapless and the degeneracy of π-edge states is partially lifted, see figure 3.2. A zoom of the dispersion relations of the π-edge states around the time-reversal invariant $\mathbf{k} = \pm 1$ point are also included in the figure 3.3. The SOI slightly shifts up in energy the midgap π-edge states, giving a small dispersion to the flat bands and lifting partially their fourfold degeneracy, except at the $\mathbf{k} = \pm 1$ point of the BZ. Therefore, the π-edge states split into two degenerated Kramers doublets with linear dispersion only in an extremely small region around the crossing $\mathbf{k}$ point and with forward and backward mover states having opposite spin. Since twofold degenerate states are confined at different edges of the ZGNR, two independent counter-propagating spin-filtered edge states are at each boundary of the ribbon. Hence, for a given energy the ZGNR has four conducting channels spatially separated, an extreme of the ribbon contains a forward mover with a given spin S and a backward mover with opposite spin, $-S$, and conversely for the other extreme of the ribbon.

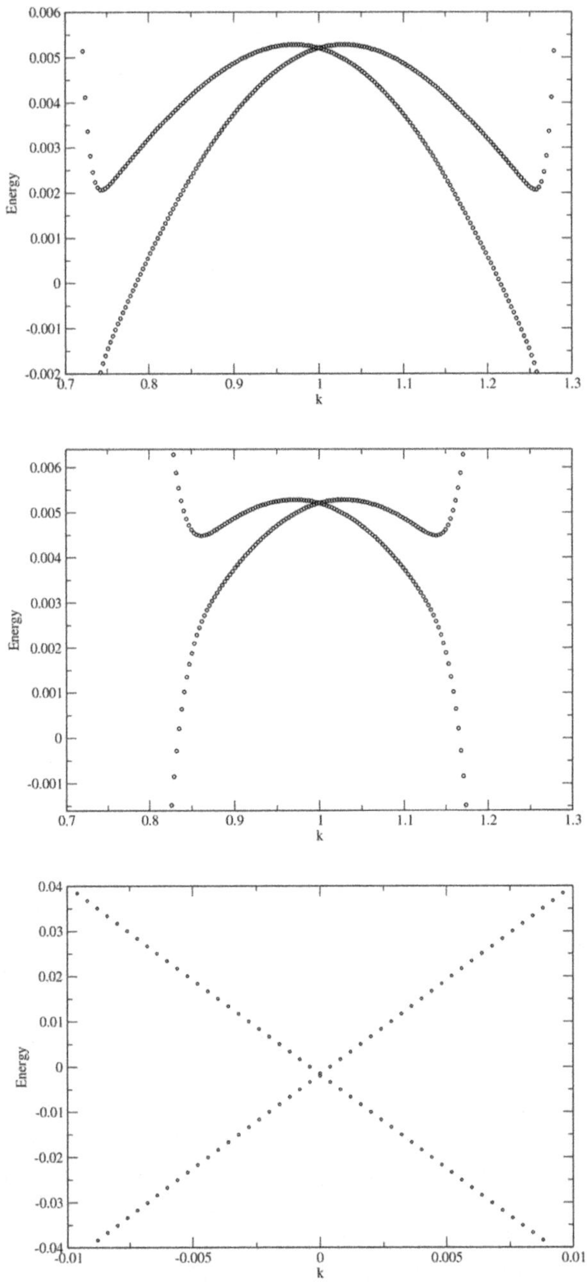

Figure 3.3. Zoom of the band structure of the zigzag ribbon $N = 60$ (top) and $N = 20$ (center) with SOI $\lambda = 0.4$ eV, for the zero-energy edge states around $\mathbf{k} = 1$. Zoom of the band structure of the armchair metallic ribbon $N = 64$ around Γ (bottom) the crossing states show the linear dispersion. k are in units of $\frac{\pi}{a}$ and energies in eV.

The midgap π-states present a localized character along the edges, their wave-function amplitude is fully localized at the edge atoms in a **k** interval around **k** = ±1. However, the SOI induces π–σ hybridization and the localization of the edge states starts to decrease as they lose their pure π-orbital character away from the high symmetry point. Therefore, the presence of the SOI and consequently of the π–σ hybridization, slightly increases the localization length of π-edge states [50]. Moreover, due to the SOI the spin degree of freedom is coupled to the lattice and $\mathbf{S_z}$ does not commute with the Hamiltonian of equation (3.1). The calculated expectation values of $<\mathbf{S}>$ shows that the spin orientation of π-edge states at the time-reversal invariant **k** = ±1 point is almost perpendicular to the graphene plane. The orientation axis, although dependent on the magnitude of k, slightly changes in the k interval in which the edge state remains spatially localized [50, 51].

Finally, the SOI splitting of the π-edge states is independent of the ZGNR width and is fully determined by the magnitude of the spin–orbit coupling strength λ. For the sake of clarity we use in the figures an unrealistically large value of 400 meV, for λ values in the range of tens of meV, the actual energy splitting of the π-states is of the order of 1.0×10^{-2} meV.

3.3.2 Armchair graphene nanoribbons

In armchair graphene nanoribbons, edges are formed by homogeneous rows of dimers with alternating atoms of the two different sublattices. The electronic properties of these ribbons present a strong dependence on their width, as it happens to zigzag CNTs. The armchair ended ribbons are metallic when its width is $W = (3m - 1)d$, where m is a positive integer and d is the C-C atom distance. For other values of W the armchair ribbons (AGNRs) are semiconductors with a direct band gap at Γ [14, 46]. The variation of the energy gap, which is inversely proportional to the ribbon width, follows three distinct family behaviours and metallic ribbons are obtained in a sequence of period 3. Moreover, all the ribbons with the armchair termination lack the existence of π-edge states. AGNR, both metallic and semiconductor, of different widths are calculated in order to clarify how the SOI affects their electronic structures.

Metallic AGNRs

The band structure of metallic AGNRs presents two states crossing at the Fermi energy at the Γ point of the BZ. In the absence of spin–orbit coupling, these crossing states are spin degenerate, have pure π-orbital character, show linear dispersion and their wave functions are extended throughout the whole ribbon width. Figure 3.4 displays the band structure of the $N = 64$ metallic AGNR. The inclusion of the SOI term in the Hamiltonian opens a small gap at Γ, for the ribbon $N = 28$ is of ≈ 1 meV for $\lambda = 400$ meV, and decreases as the width of the ribbon increases [52], for the same value of λ, the gap is of $\approx 10^{-1}$ meV for a ribbon of $N = 64$. But the two states sustain the linear dispersion, remain twofold degenerate and keep its extended character, see right panel of figure 3.3. SOI slightly changes the orbital composition of the linear states allowing for a small proportion of σ-orbitals and the pure spin

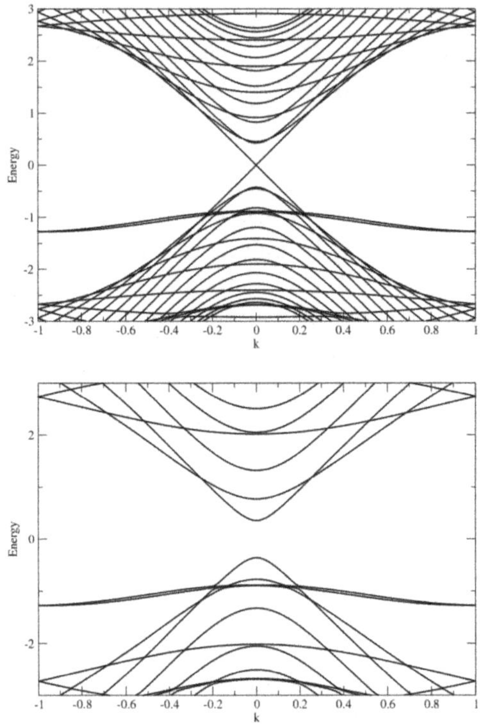

Figure 3.4. Band structure of the armchair $N = 64$ metallic ribbon (top) and $N = 24$ semiconductor ribbon (bottom) with SOI $\lambda = 0.4$ eV. k are in units of $\frac{\pi}{a}$ and energies in eV.

nature is not conserved. It has been found that, within the low-energy Kane-Mele model [53], these states persist in the presence of intrinsic SOI as spin-filtered states located at the ribbon edges [54]. However, in the four-orbital TB model to obtain location of these states at the edges of the ribbon, either, unrealistic high values of λ, -almost of the same order of magnitude that the intersite hopping- or extremely wide ribbons, are needed [50, 51, 54].

Conversely, in AGNR with unsaturated bonds at the borders, in the absence of SOI, edge states composed mainly of σ-orbitals and fourfold degenerated appear below the Fermi level. They lie below -1 eV at the $\mathbf{k} = \pm 1$ points of the BZ, are moderately dispersive and fully localized at the edge atoms of the ribbon. Around $\mathbf{k} \approx \pm 0.3$ they merge with the bulk bands, but keep its fourfold degeneracy and localized character. Their wave-function amplitude is confined to the outermost atoms at each border. These σ-orbital derived states are truly surface states appearing at the same energy independently of the ribbon width, as we have confirmed calculating AGNRs of up to $n = 122$ and $n = 242$. They are equivalent to the π- and σ-edge states present in ZGNR and analogously to them upon SOI inclusion, split into two spin-filtered Kramers doublets. They keep the localization at the boundary atoms and although their orbital composition presents now a small proportion of π orbitals, they have a defined spin. The so-called bulk subbands of the ribbon always maintain their double degeneracy, because the inversion symmetry of the AGNR.

Semiconductor armchair GNRs
The effects of the SOI in the band structure of the semiconductor AGNRs are similar to those described in the metallic ribbons. The energy gap slightly changes by the inclusion of the SOI. There are energy shifts of the valence occupied and empty conduction bands but they move in opposite energy direction. The energy shifts are of the same order of magnitude than those obtained in the metallic AGNRs. Twofold degeneracy is maintained for all the bands and those around the Fermi level remains extended across the ribbon, see bottom panel of figure 3.4.

Consequently, our results indicate that in order to have spin-filtered edge states, using sizable values of the SOI coupling constant, the pre-existence of surface states is an essential requirement.

3.3.3 Curvature effects in GNRs

Curved GNRs has been obtained by opening CNTs, therefore it seems interesting to investigate the curvature effects in the GNRs properties. Curvature is introduced by bending the ribbon along its width, while the bonding distances, and therefore the hoppings, between atoms in the edge direction are not modified. The bending is realized without stretching, varying the atom coordinates to form and open cylinder. Different curvatures are obtained by changing the angles and diameters of the cylindrical configurations.

Curvature induces hybridization between π- and σ-orbitals and thus it amplifies the SOI effects in CNTs. However, our results show that curvature effects are weaker for ribbons than for CNTs, due to the different symmetry of the two 1D-graphene structures [52, 55]. In ZGNRs, in the absence of SOI, the behaviour of π-edge states is affected by curvature. They appear at zero-energy as in the flat ribbon, but present a small dispersion and their orbital composition is slightly changed with a small admission of σ-orbitals in the curved ribbon. Hence, in narrow curved ribbons π-states localized at the borders interact and accumulate at both edges. In the presence of SOI, curvature induces a slight increase of the energy splittings in ZGNRs and a small variation of the gap in AGNRs, with respect to those of the flat ribbons. Consequently, the major effect is on the degenerate π-edge states, while in a flat ribbon they become spin-filtered by the SOI almost independently of the width of the ribbon, in narrow curved ribbons, the interaction between states at both edges induced by curvature, prevents the formation spin conducting channels. Only for wide ribbons the edge state interaction induced by curvature is negligible.

Furthermore, curvature results in a variation of the spin orientation axis and in a significant increase of the orbital angular momentum. The expectation value of $<S>$ for the π-edge states is no longer perpendicular to the graphene sheet, but an in-plane contribution turns the spin axis to form an angle with the ribbon plane. In addition, $<L>$ increases by more than two orders of magnitude with respect to that of the corresponding flat ribbon [50, 51].

3.4 Summary

We have investigated the effects of the intrinsic SOI in graphene nanoribbons of different edge-shapes and widths and analyzed the existence of spin-filtered edge states at the boundaries. We describe graphene by a four-orbital -$2s$, $2p_x$, $2p_y$ and $2p_z$-TB model, which takes into account not only the discreteness of the lattice but also the σ-orbitals forming the C-C in-plane bonds. Within this model the on-site approach for the intrinsic SOI couples p-orbitals on the same atom.

We found a subtle correlation between SOI and the geometry of the ribbon. Since the GNRs, both ZGNR and AGNR, have spatial inversion symmetry all the electronic states, either edge or bulk related, are twofold degenerated. Even the interplay between curvature and SOI does not break the degeneracy. In general, the effect of SOI is to induce π–σ hybridization, reduce the localization of 1D edge states and deviate the spin-orientation from the perpendicular direction to the graphene sheet. However, when edge states exits even in the absence of SOI, such as the π-states of ZGNR, the SOI turns edge state into helical 1D-states, spin-filtered and localized at the extreme of the ribbons. Conversely, if there is not edge states at the borders, as it happens in AGNR, the magnitude of the SOI needed to induce a topological phase transition and turn graphene into a Quantum Spin Hall Insulator (QSHI) is extremely large, almost of the same order of magnitude that the intersite hopping. Unphysical values of the order of tenths of eV are still unable to induce the topological transition.

Consequently, while the one-orbital Kane-Mele model predicts that intrinsic SOI alone would turn graphene into a QSHI, in the four-orbital TB model intrinsic SOI is unable to induce the topological transition. In the Kane-Mele model the SOI opens a valley- and spin-dependent gap of the order of the SOI coupling constant, therefore making the ribbon sufficiently wide the confinement gap is smaller that the SOI gap and the topological phase transition occurs. In the four-orbital model, the SOI splittings and shifts are not linear in the coupling constant, and they are at least two orders of magnitude smaller. Therefore, the topological transition requires a physically unrealistic scenario. Nevertheless, for GNR with edge geometries supporting edge states, the SOI leads to spin-filtered 1D-helical sates fully localized at the edges, similar to those resulting from the bulk-surface correspondence of the QSHI phase.

Acknowledgments

The authors thank J.I. Beltrán for fruitful conversations. MPLS acknowledges financial support of Spanish MINECO under grant No. PGC2018–099199-B and the European Union structural funds and the Comunidad Autónoma de Madrid (CAM) NMAT2D-CM Program (S2018-NMT-4511). MCM acknowledges financial support of MINECO and FEDER Program of the European Union, MAT2015–6688-C3–1-R and RTI2018-097895-B-C41.

References

[1] Kane C L and Mele E J 2005 Quantum spin Hall effect in graphene *Phys. Rev. Lett.* **95** 226801

[2] Kane C L and Mele E J 2005 Z_2 topological order and the quantum spin Hall effect *Phys. Rev. Lett.* **95** 146802

[3] Hasan M Z and Kane C L 2010 Colloquium: topological insulators *Rev. Mod. Phys.* **82** 3045–67

[4] Castro Neto A H, Guinea F, Peres N M R, Novoselov K S and Geim A K 2009 The electronic properties of graphene *Rev. Mod. Phys.* **81** 109–62

[5] Zhang Y, Tan Y W, Stormer H and Kim P 2005 Experimental observation of the quantum Hall effect and Berry phase in grapheme *Nature* **438** 201–4

[6] Bolotin K I, Ghahari F, Shulman M D, Stormer H and Kim P 2009 Observation of the fractional quantum Hall effect in grapheme *Nature* **462** 196–9

[7] Persin D and MacDonald A H 2012 Spintronics and pseudospintronics in graphene and topological insulators *Nat. Mater.* **11** 409–16

[8] Trauzettel B, Bulaev D V, Loss D and Bukard G 2007 Spin qubits in graphene quantum dots *Nat. Phys.* **3** 192–6

[9] Jiao L, Zhang L, Wang X, Diankov G and Dai H 2009 Narrow graphene nanoribbons from carbon nanotubes *Nature* **458** 877–80

[10] Jia X *et al* 2009 Controlled formation of sharp zigzag and armchair edges in graphitic nanoribbons *Science* **323** 1701–05

[11] Cai J *et al* 2010 Atomically precise bottom-up fabrication of graphene nanoribbons *Nature* **466** 470–3

[12] Ruffieux P *et al* 2016 On-surface synthesis of graphene nanoribbons with zigzag edge topology *Nature* **531** 489–92

[13] Narita A *et al* 2014 Synthesis of structurally well-defined and liquid-phase-processable graphene nanoribbons *Nat. Chem.* **6** 126–32

[14] Son Y-W, Cohen M L and Louie S G 2006 Energy gaps in graphene nanoribbons *Phys. Rev. Lett.* **97** 216803

[15] Cao T, Zhao F and Louie S G 2017 Topological phases in graphene nanoribbons: junction states, spin centers, and quantum spin chains *Phys. Rev. Lett.* **119** 076401

[16] Zak J 1989 Berry's phase for energy bands in solids *Phys. Rev. Lett.* **62** 2747–50

[17] Gröning O *et al* 2018 Engineering of robust topological quantum phases in graphene nanoribbons *Nature* **560** 209–13

[18] Rizzo D J *et al* 2018 Topological band engineering of graphene nanoribbons *Nature* **560** 204–8

[19] Lee Y-L, Zhao F, Cao T, Ihm J and Louie S G 2018 Topological phases in cove-edged and chevron graphene nanoribbons: geometric structures, Z_2 invariants, and junction states *Nano Lett.* **18** 7247–53

[20] Lin K-S and Chou M-Y 2018 Topological properties of gapped graphene nanoribbons with spatial symmetries *Nano Lett.* **18** 7254–60

[21] Ando T 2000 Spin-orbit interaction in carbon nanotubes *J. Phys. Soc. Jpn.* **69** 1757

[22] Chico L, López-Sancho M P and Muñoz M C 2004 Spin splitting induced by spin-orbit interaction in chiral nanotubes *Phys. Rev. Lett.* **93** 176402

[23] De Martino A, Eggert R, Hallberg K and Balseiro C A 2002 Spin-orbit coupling and electron spin resonance theory for carbon nanotubes *Phys. Rev. Lett.* **88** 206402

De Martino A, Eggert R, Hallberg K and Balseiro C A 2004 Spin-orbit coupling and electron spin resonance for interacting electrons in carbon nanotubes *J. Phys.:Condens. Matter* **16** S1437

[24] Jespersen T S *et al* 2011 Gate-dependent spin–orbit coupling in multielectron carbon nanotubes *Nat. Phys.* **7** 348–53

[25] Steele G A *et al* 2013 Large spin–orbit coupling in carbon nanotubes *Nat. Commun.* **4** 1573

[26] Churchill H O H, Kuemmeth F, Harlow J W, Bestwick A J, Rashba E I, Flensberg K, Stwertka C H, Taychatanapat T, Watson S K and Marcus C M 2009 Relaxation and dephasing in a two-electron ^{13}**C** nanotube double quantum dot *Phys. Rev. Lett.* **102** 166802

[27] Jhang S H, Marganska M, Skourski Y, Preusche D, Witkamp B, Grifoni M, van der Zant H, Wosnitza J and Strunk C 2010 Spin–orbit interaction in chiral carbon nanotubes probed in pulsed magnetic fields *Phys. Rev.* B **82** 041404

[28] Kuemmeth F, Ilani S, Ralph D C and McEuen P L 2008 Coupling of spin and orbital motion of electrons in carbon nanotubes *Nature* **452** 448

[29] Chico L, López-Sancho M P and Muñoz M C 2009 Curvature-induced anisotropic spin–orbit splitting in carbon nanotubes *Phys. Rev.* B **79** 235423

[30] Izumida W, Sato K and Saito R 2009 Spin-orbit interaction in single wall carbon nanotubes: symmetry adapted tight-binding calculation and effective model analysis *J. Phys. Soc. Jpn.* **78** 074707

[31] Jeong J-S and Lee H-W 2009 Curvature-enhanced spin–orbit coupling in a carbon nanotube *Phys. Rev.* B **80** 075409

[32] Sichau J, Prada M, Anlauf T, Lyon T J, Bosnjak B, Tiemann L and Blick R H 2019 Resonance microwave measurements of an intrinsic spin–orbit coupling gap in graphene: a possible indication of a topological state *Phys. Rev. Lett.* **122** 046403

[33] Wang Z *et al* 2015 Strong interface-induced spin–orbit interaction in graphene on WS_2 *Nat. Commun.* **6** 8339

[34] Brey L 2015 Spin–orbit coupling in graphene induced by adatoms with outer-shell *p* orbitals *Phys. Rev.* B **92** 235444

[35] Weeks C, Hu J, Alicea J, Franz M and Wu R 2011 Engineering a robust quantum spin Hall state in graphene via adatom deposition *Phys. Rev.* X **1** 021001

[36] Huertas-Hernando D, Guinea F and Brataas A 2006 Spin-orbit coupling in curved graphene, fullerenes, nanotubes, and nanotube caps *Phys. Rev.* B **74** 155426

[37] Chadi D J 1977 Spin-orbit splitting in crystalline and compositionally disordered semiconductors *Phys. Rev.* B **16** 790

[38] Gallego S and Muñoz M C 1999 Spin-orbit effects in the surface electronic configuration of the $Pt_3Mn(111)$ and layered (2x") $Pt/Pt_3Mn(111)$ compounds *Surface Sci.* **423** 324

[39] Tománek D and Louie S G 1988 First-principles calculation of highly asymmetric structure in scanning-tunneling-microscopy images of graphite *Phys. Rev.* B **37** 8327

[40] Gmitra M, Konschuh S, Ertler C, Ambrosch-Draxl C and Fabian J 2009 Band-structure topologies of graphene: spin–orbit coupling effects from first principles *Phys. Rev.* B **80** 235431

[41] Min H, Hill J E, Sinitsyn N A, Sahu B R, Kleinman L and MacDonald A H 2006 Intrinsic and rashba spin–orbit interactions in graphene sheets *Phys. Rev.* B **74** 165310

[42] Cysne T P, Ferreira A and Rappoport T G 2018 Crystal-field effects in graphene with interface-induced spin–orbit coupling *Phys. Rev.* B **98** 045407

[43] Gmitra M and Fabian J 2015 Graphene on transition-metal dichalcogenides: a platform for proximity spin–orbit physics and optospintronics *Phys. Rev.* B **92** 155403

[44] Gmitra M, Kochan D, Högl P and Fabian J 2016 Trivial and inverted Dirac bands and the emergence of quantum spin Hall states in graphene on transition-metal dichalcogenides *Phys. Rev.* B **93** 155104

[45] Brey L and Fertig H A 2006 Electronic states of graphene nanoribbons studied with the Dirac equation *Phys. Rev.* B **73** 235411

[46] Motohiko E 2006 Peculiar width dependence of the electronic properties of carbon nanoribbons *Phys. Rev.* B **73** 045432

[47] Fujita M, Wakabayashi K, Nakada K and Kusakabe K 1996 Peculiar localized state at zigzag graphite edge *J. Phys. Soc. Jpn.* **65** 1920

[48] Peres N M R, Castro Neto A H and Guinea F 2006 Conductance quantization in mesoscopic graphene *Phys. Rev.* B **73** 195411

[49] Ryu S and Hatsugai Y 2002 Topological origin of zero-energy edge states in particle-hole symmetric systems *Phys. Rev. Lett.* **89** 077002

[50] López-Sancho M P and Muñoz M C 2011 Intrinsic spin-orbit interactions in flat and curved graphene nanoribbons *Phys. Rev.* B **83** 075406

[51] Gosálbez-Martnez D, Palacios J J and Fernández-Rossier J 2011 Spin–orbit interaction in curved graphene ribbons *Phys. Rev.* B **83** 115436

[52] Onari S, Ishikawa Y, Kontani H and Inoue J-I 2008 Intrinsic spin Hall effect in graphene: numerical calculations in a multiorbital model *Phys. Rev.* B **78** 121403

[53] Zarea M and Sandler N 2007 Electron-electron and spin–orbit interactions in armchair graphene ribbons *Phys. Rev. Lett.* **99** 256804

[54] San-Jose P, Brey L, Fertig H A and Prada E 2011 Band topology and quantum spin hall effect in bilayer graphene *Solid State Commun.* **151** 1075

[55] de Juan F, Cortijo A and Vozmediano M A H 2007 Charge inhomogeneities due to smooth ripples in graphene sheets *Phys. Rev.* B **76** 165409

IOP Publishing

Graphene Nanoribbons

Luis Brey, Pierre Seneor and Antonio Tejeda

Chapter 4

Emergent quantum matter in graphene nanoribbons

J L Lado, R Ortiz and J Fernández-Rossier

4.1 Introduction

In this chapter, we provide a perspective on the potential of graphene nanoribbons as a platform to host a variety of non-trivial emergent electronic states, such as topological phases, quantum spin liquids, broken symmetry magnetic states and Yu–Shiba–Rusinov excitations. This potential arises from the capability to nano-engineer the electronic properties of nanoribbons using several different resources:

- Geometrical control. Graphene ribbons with different shapes, orientation, widths, can be synthesized [1]. This gives rise to different electronic properties, including the emergence of localized edge and interface states that can host unpaired spin electrons.
- Tuning the electron density. Using either gating and chemical doping, it is possible to control the density of electrons.
- The electronic properties of GNR can be affected by several types of proximity effect: spin–orbit, superconducting and magnetic. Therefore, they provide a unique platform to explore the interplay between local magnetic order and superconductivity.

In addition, scanning probe spectroscopies permit one to probe both the structural properties of GNR as well as their electronic properties [2–5] and spin excitations [6], with atomic scale resolution, and constitute a great tool to probe the emergent electronic phases [7].

The most compelling argument to expect non-trivial correlated phases in GNR comes from experiments. Non-trivial phases, including Mott–Hubbard insulating phases [8] and non-trivial superconductivity [9], have been observed in twisted graphene bilayers. Whereas the precise origin of the superconductivity is not understood, there is a consensus on the crucial role played by an array of localized

doi:10.1088/978-0-7503-1701-6ch4 © IOP Publishing Ltd 2020

states that form very narrow bands at the Dirac energy. Given that the chemical properties of monolayer GNR are almost identical to those of graphene bilayer, it is our contention that the localized zero modes of the GNR can also result in non-trivial correlated phases. The scope of this chapter is to unveil some physical mechanisms that can promote non-trivial electronic phases, rather than the technical aspects of how to model them.

We focus on three classes of non-trivial electronic behavior. First, we revisit the thoroughly studied problem of magnetic order in the zigzag edges. We address the prominent role of spin fluctuations in this low dimensional system. Second, we discuss how to realize spin chain Hamiltonians in graphene ribbons engineered to host an ordered array of localized zero mode states. Third, we address the interplay between emergent local moments and superconducting proximity effect.

4.2 Modeling GNR

4.2.1 Geometries

In this chapter, we focus on GNR with atomically precise edges along the crystallographic axis of graphene. These are the so-called armchair and zigzag edges. We consider both finite (0D) and infinite (1D) ribbons. We also consider ribbons with a periodic modulation of their width, that are known to host interfacial topological zero modes [10–12]. Several other geometries, such as chiral ribbons with sufficiently long zigzag patches, as well as chevron type ribbons with zigzag edges, can result in the formation of local moments and non-trivial spin physics.

4.2.2 Single particle terms

Here, we adopt the standard model [13, 14] to describe GNR, namely, a single orbital tight-binding model with first neighbor hopping t. The single orbital is the p_z atomic orbital of carbon, that is decoupled from the rest in planar structures. Unless otherwise stated, we assume that edge carbon atoms are passivated with hydrogen, so that there are no dangling bonds at the surface. This model gives a fair description of the states in a few eV window around the Fermi energy for planar carbon based structures, going from zero dimensional molecules, to planar graphene. It is standard [13] to take $t = -2.7$ eV, which provides a good slope for the Dirac cones in planar graphene. The first neighbor hopping Hamiltonian reads:

$$\mathcal{H}_0 = -t \sum_{i,i',\sigma} c_{i\sigma}^{\dagger} c_{i',\sigma} \tag{4.1}$$

where i' stand for the first neighbors of i. This is the dominant term in the Hamiltonian and it accounts for the Dirac cones in graphene [13, 14], the existence of localized states in the zigzag edges [15], and the gapped bands in armchair GNR [15].

In addition, we consider the effect of the several spin-dependent terms in the Hamiltonian. First, the Zeeman term, given by

$$\mathcal{H}_Z = \frac{1}{2} g\mu_B \vec{B} \cdot \sum_{i,\sigma,\sigma'} \vec{\sigma}_{\sigma,\sigma'} c_{i\sigma}^{\dagger} c_{i,\sigma'} \tag{4.2}$$

where $g \simeq 2$ and $\vec{\sigma}_{\sigma,\sigma'}$ are the Pauli matrices.

The intrinsic spin–orbit coupling, proposed by Kane and Mele, is described by [16]:

$$H_{KM} = \sum_{i,i''\sigma} it_{KM}\sigma\nu_{i,i''} c_{i\sigma}^{\dagger} c_{i''\sigma} \tag{4.3}$$

where i'' stands for the second neighbors of i, summation, $\sigma = \pm 1$ are the spin projections (along the axis perpendicular to the crystal plane) and $\nu_{i,i''} = +(-)1$ for clockwise (anticlockwise) second neighbor hopping. When added to the hopping Hamiltonian (4.1), the Kane-Mele term opens a topologically non-trivial band-gap $\Delta_{SOC} = 6\sqrt{3}\, t_{KM}$ at the Dirac points. The non-trivial nature of the gap implies the emergence of spin-locked chiral edge states [16]. Because of the small magnitude of Δ_{SOC} in graphene, the observation of this gap is very challenging [17] and the localization length of the edge states is very large. Therefore, this term has a minor influence in the properties of graphene ribbons. However, this type of term could be enhanced by proximity effect [18].

A second type of spin obit effect can arise when mirror symmetry is broken, due to the application of an external off-plane electric field, or due to interaction with the substrate. This is the so-called Rashba spin–orbit term [16, 19]:

$$\mathcal{H}_R = it_R \sum_{i,j,s,s'} \vec{E} \cdot \left(\vec{r}_{ij} \times \vec{\sigma}\right)_{s,s'} c_{is}^{\dagger} c_{js'} \tag{4.4}$$

where $\vec{r}_{ij}$ is unit vector along the bond between the carbon sites i and j, $\vec{\sigma}$ are the spin Pauli matrices and $\vec{E}$ is a vector related to inversion symmetry breaking of the graphene lattice, such as an off-plane electric field [11, 19]. The Rashba spin–orbit coupling does not commute with S_z and promotes mixing between the two spin channels.

4.2.3 Coulomb interaction

In this chapter, we consider the effect of electron-electron Coulomb interactions within the Hubbard approximation:

$$\mathcal{H}_U = U \sum_i n_{i\uparrow} n_{i\downarrow} \tag{4.5}$$

where U stands for the Coulomb penalty for having two electrons in the same π orbital in a single carbon atom. The value of U may depend on additional screening effects, including the substrate. In addition, the right value of U might depend on whether or not we include a next-neighbor Coulomb repulsion in the Hamiltonian [20]. Here, we adopt U as a variable parameter that takes values in the range of $U = |t|$.

The Hubbard model can only be solved exactly in very specific geometries, such as the monostrand one-dimensional chain. Thus, very often [21–27] the model is treated at the mean field approximation, where the exact Hamiltonian is replaced by an effective Hamiltonian

$$\mathcal{H}_{U,MF} = \mathcal{H}_{\text{Hartree}} + \mathcal{H}_{\text{Fock}} \tag{4.6}$$

where

$$\mathcal{H}_{\text{Hartree}} = U(n_{i,\uparrow}\langle n_{i,\downarrow}\rangle + n_{i,\downarrow}\langle n_{i,\uparrow}\rangle) \tag{4.7}$$

$$\mathcal{H}_{\text{Fock}} = -U\left(c^{\dagger}_{i,\downarrow}c_{i,\uparrow}\left\langle c^{\dagger}_{i,\uparrow}c_{i,\downarrow}\right\rangle + c^{\dagger}_{i,\uparrow}c_{i,\downarrow}\left\langle c^{\dagger}_{i,\downarrow}c_{i,\uparrow}\right\rangle\right) \tag{4.8}$$

so that electrons interact with an external field that is self-consistently calculated. Most often [21–24, 28–30], an additional approximation has been used that assumes a collinear magnetization so that the Fock term vanishes. For small nanographenes, such as triangular and hexagonal islands with zigzag edges [22], the results of collinear mean-field calculations of the Hubbard model are very similar to those obtained using density functional theory calculations that include long-range Coulomb interactions and include several atomic orbitals per carbon atom. The same statement holds true for infinitely long graphene ribbons with zigzag edges: both mean field calculations Hubbard model calculations [21, 23] and DFT based calculations [31] predict ferromagnetic order at the edges and antiferromagnetic inter-edge coupling at half filling.

The study of non-collinear magnetization has permitted us to study the canted spin phases in graphene quantum Hall bars [26], as well as the existence of in-gap topological fractional excitations at the domain walls of graphene zigzag ribbons [27].

4.2.4 Proximity terms

The effective Hamiltonian for electrons in graphene can be modified due to the interaction with the substrate. The most frequently considered types of proximity terms are a sublattice symmetry-breaking on-site potential, which opens up a gap [30, 32], a ferromagnetic spin proximity effect, that splits the bands [33–36] and a superconducting proximity effect that adds a pairing term to the Hamiltonian, and opens up a superconducting gap to graphene whenever the Fermi energy lies on a band.

The on-site potential can be written down as:

$$H_J = \sum_i W(i) c^{\dagger}_{i\sigma} c_{i\sigma}. \tag{4.9}$$

Whenever $W(i)$ is different for A and B sublattice, this term can open up a gap in graphene. When the sign of this gap is modulated across graphene, kink states can emerge [37].

The spin proximity effect can be written down as:

$$H_J = \frac{1}{2}\sum_i \vec{J}(i) \cdot \vec{\sigma}_{\sigma,\sigma'} c^{\dagger}_{i\sigma} c_{i\sigma'} \tag{4.10}$$

where $\vec{J}(i)$ is the exchange field that is proportional to the magnetization field of the proximity layer. In the simplest scenario, this is taken as a collinear and constant field, so that the spin proximity effect leads to a spin splitting of the bands that, in conjunction with Rashba spin–orbit coupling, can induce a quantized anomalous Hall phase [33]. Spin proximity with non-collinear or even non-coplanar substrates, such as skyrmions, can also result in a quantized anomalous Hall phase, without the need of spin–orbit coupling [38]. The typical magnitude for the exchange splitting, as obtained from DFT calculations [34–36], is in the range of a few tens of meV at most. Experimentally, a report of splitting induced by spin proximity effect observed in graphene is much smaller than that in the range of a fraction of meV [39].

The superconducting proximity effect is introduced as an effective conventional s-wave pairing term:

$$H_{SC} = \Delta \sum_i \left[c_{i,\uparrow} c_{i,\downarrow} + c^{\dagger}_{i,\downarrow} c^{\dagger}_{i,\uparrow} \right]. \tag{4.11}$$

This term has to be treated using the so-called Bogoliubov–de Gennes (BdG) Hamiltonian [40].

4.3 Emergent phases and zero modes

4.3.1 Single particle theory of zero modes

We consider graphene ribbons not too far from their charge neutrality point. Therefore, their Fermi energy lies close to the Dirac point. Because of quantum confinement, extended states of graphene ribbons are gapped. This leads to semi-conducting or insulating ribbons that are not expected to host non-trivial electronic phases. The way out of this situation comes from the existence of zero modes. In pristine GNR, zero modes arise in the following instances:

1. At sufficiently long zigzag edges [15]. As we discuss below, there is one zero mode for every three carbon atoms in a zigzag edge.
2. At interfaces between gapped armchair ribbons with different symmetry protected topological indexes Z_2, defined below, as proposed by Cao *et al* [10]

In addition, zero modes also appear when graphene is functionalized with atomic hydrogen [41–44] or any other sp^3 functionalization [45].

There are two complementary ways to understand the emergence of these zero modes. The first way invokes the bipartite character of the honeycomb lattice and the emergence of at least $N_A - N_B$ zero modes [42, 46], where N_A and N_B are the number of sites in the two sublattices that form a structure. In addition, the theorem permits to anticipate the sublattice polarized nature of the zero modes. This first method permits prediction of the emergence of zero modes in sp^3 functionalized graphene [42]. There, the p_z orbital forms a strong covalent bond with an orbital of the functionalizing species, such as the 1*s* orbital of atomic hydrogen. This takes away both one electron and one orbital from the p_z array. This can be effectively modeled as a tight-binding model with a missing site [42, 47]. The sublattice imbalance argument can also be applied right away to the interface states between armchair ribbons [11], shown in figure 4.1. In the case of graphene zigzag edges, it can be invoked, although in a less rigorous manner. Locally, zigzag edges have sublattice imbalance, but globally, the structures have $N_A = N_B$.

The existence of zero modes in some GNR can also be related to topological arguments. The interface between two media described with different topological indexes, N_1 and N_2, is expected to host at least $N_1 - N_2$ zero modes. In a 1D crystal with mirror and inversion symmetry, we define the Zak phase of a band *n* as [10, 48]:

$$\gamma_n = i\left(\frac{2\pi}{d}\right)\int_{-\pi/d}^{\pi/d} dk \left\langle u_{nk} | \frac{\partial u_{nk}}{\partial k} \right\rangle \tag{4.12}$$

where *d* is the unit cell size and u_{nk} is the periodic part of the Bloch wave function for band *n*. In a symmetry protected 1D crystal, the Zak phase is quantized as 0 or π modulo 2π. This permits us to define a topological index:

$$(-1)^{Z_2} = e^{i\sum_n \gamma_n}. \tag{4.13}$$

From the bulk-boundary correspondence, symmetric junctions of armchair ribbon with different Z2 numbers are expected to host localized zero modes at the interfaces. This has been confirmed both with DFT [10] and tight-binding calculations [11]. These junctions happen to have $|N_A - N_B| = 1$, so that the interface zero mode can be understood using the theorem for bipartite lattice. Using similar

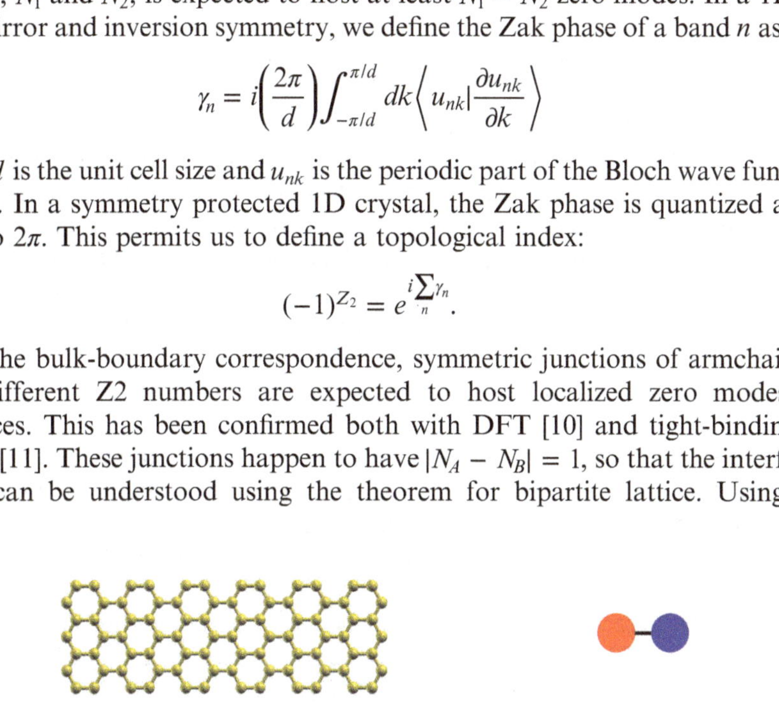

Figure 4.1. Left column: different GNRs considered in this chapter. Top: a finite size graphene ribbon. Middle: 1D graphene ribbon that alternates sections with different widths and armchair edges. Bottom: a 1D ribbon with zigzag edges. Right column: the equivalent lattice spin model.

arguments [48], the existence of edge modes in 1D zigzag edges has been related to the Zak phase for the family of 1D states defined in a cut of the 2D Brillouin zone.

4.3.2 Infinite ribbons

We begin our discussion of specific systems with the case of one-dimensional graphene ribbon with zigzag edges. As shown in figure 4.2(a), the energy bands feature two flat bands at $E = 0$. These two bands of zero modes occupy exactly one third of the Brillouin zone. Given that the unit cell of the ZZ GNR has exactly one carbon site per edge, this implies that the ratio of zero modes per carbon edge atom is 1/3. The wave function of the edge modes is sublattice polarized, and its amplitude quickly decays as we move inwards in the GNR. Other than these zero modes, the rest of the bands are gapped, reflecting the confinement of the Dirac particles in the section of the ribbon.

The flat bands at $E = 0$ give rise to a very large density of states at that energy. Given that $E = 0$ is the Fermi energy for half filling, interactions are expected to have a strong impact in this system. This was found out more before the turn of the century, using a mean field approximation for the Hubbard model in this system [21] and subsequent work, using both the Hubbard model [23–26] and DFT calculations. In all instances, the predictions of these symmetry breaking methods are:

- The zigzag edges are ferrromagnetic, with magnetic moments in the range of 0.15 μ_B per carbon atom.
- The inter-edge interaction is antiferromagnetic and decays rapidly as a function of the width.
- The energy bands show a dispersion of the edge states, driven by the interactions [23]. In the case of parallel (antiparallel) alignment of the edge magnetizations, the ZZ is a conductor (insulator). This finding prompted proposals for using graphene ribbons as ideal spin valves [24, 49].

The qualitative effect of interactions can be captured by adding local exchange fields at the upper and lower zigzag edge, giving rise to results comparable to the full self-consistent calculation (figures 4.2(b) and (c)).

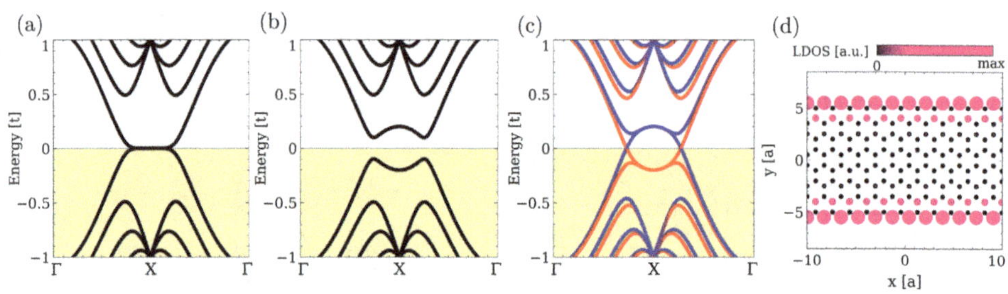

Figure 4.2. (a) Band structure of a zigzag nanoribbon with a 16 atom width. Band structure with local exchange fields in the upper and lower edges, aligned antiferromagnetically (b) and ferromagnetically (c). (d) Spatial distribution of the flat band of (a), which becomes magnetized in (b), (c).

In general, mean-field calculations for any nanographene with zigzag edges predict the existence of magnetic moments localized at the edges with ferromagnetic correlations between edges that belong to the same sublattice, and antiferromagnetic correlations between edges that belong to opposite sublattices [22]. In the case of infinite 1D ribbons, these calculations have an obvious problem: they predict an infinitely long range order along the edge, breaking a continuous symmetry in one dimension. This is incompatible with well established theorems. In one dimension, quantum fluctuations are known to destroy this type of long range order. Therefore, we need to carry out a treatment that models this system without this drawback. Before doing that, a possible way out would be to include the terms in the Hamiltonian that break the SU(2) spin symmetry, given that 1D order is possible at $T = 0$ in Ising chains, for instance. It has been shown [25] that intrinsic spin–orbit coupling favors in-plane edge magnetization. As a result, the group of symmetry is reduced, but is still a continuous O(2) symmetry, for which no long range order can exist in 1D. In addition, the value of the magnetic anisotropy scales with the square of the intrinsic spin–orbit coupling term in the Kane–Mele Hamiltonian, which is in the range of a few tens of μeV in graphene [19]. Therefore, the magnetic anisotropy driven by the intrinsic spin–orbit coupling in graphene is negligible.

The effect of spin wave fluctuations was considered by Yazyev and Katnelson [50]. They computed the spin correlation functions along the edges using a spin ladder model of an infinite ribbon and found a power law decay, with temperature dependence spin correlation length. A quantum Monte Carlo description for the same spin ladder model for zGNR was also carried out [51]. Beyond the mean field explorations of edge ferromagnetism in zGNR have also been addressed with fermionic models including long range Coulomb interactions using both exact diagonalizations in a restricted active space in the reciprocal state [52] as well as quantum Monte Carlo simulation [53]. Both methods confirm intra-edge ferromagnetic correlations. In any event, it is apparent that a rigorous quantum theory for the edge magnetism has to go beyond broken symmetry solutions in order to include a proper treatment of quantum fluctuations.

4.3.3 Finite ribbons

We now consider two different finite size ribbons (see figure 4.3). Both of them have two weakly hybridized zero modes. The first one is a ribbon with two long armchair edges and two short zigzag edges, shown in figure 4.3, which host just one edge state each. The second structure combines armchair ribbons with different widths and mirror symmetries, such that the interface hosts zero modes. For these structures, we take periodic boundary conditions so that there are no free zigzag edges. In both cases, the structures have two zero modes inside a quite large gap. At half filling, the two edge modes host one electron. We can treat these systems, including interactions, by considering configurations where the valence state are doubly occupied and the conduction states are empty. Therefore, we have a problem of two electrons in two sites that can be solved analytically [11]. In both structures, we can change the

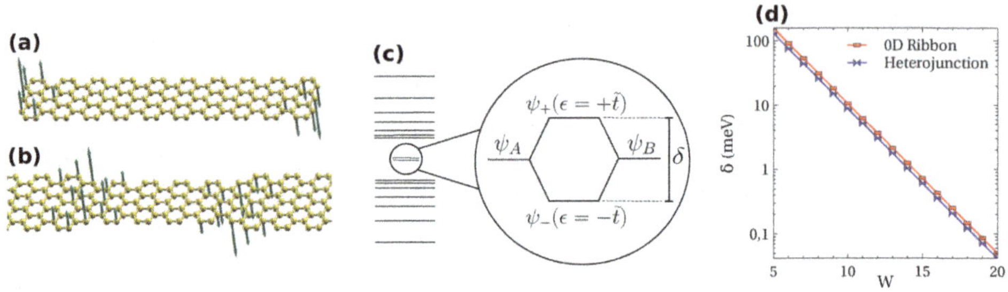

Figure 4.3. Left: GNR with short zigzag edges (top) and aGNR with two widths. Both structures have in-gap states that host local moments when Coulomb interactions are included. The vector maps reflect the magnetization obtained in a mean field calculation with the Hubbard model. Right: scheme of the single-particle $U = 0$ energy spectrum showing two almost degenerate in-gap states. Their wave functions are the symmetric and antisymmetric combinations of the sublattice polarized zero modes located at the edges/interfaces. The exponential dependence of the hybridization, as measured by the in-gap splitting, as a function of W, the length scale that controls the size of the GNR.

dimensions of the system, and thereby W, defined as the fact that the distance between either the edge or the interface controls the hybridization of the zero modes.

The non-interacting spectrum. A scheme of the single-particle spectrum characteristic of these gapped 0D GNR with two in-gap states is shown in figure 4.3(b). The energies and wave-functions of the in-gap states are denoted by $\varepsilon_\pm$ and $\psi_\pm$, respectively. It is always possible [11] to write down the wave function of a couple of conjugate states, with single-particle energy E and $-E$, in terms of the same sublattice polarized states ψ_A and ψ_B. Therefore, we write

$$\begin{aligned} \psi_A(i) &\equiv \frac{1}{\sqrt{2}}\big(\psi_+(i) + \psi_-(i)\big) \\ \psi_B(i) &\equiv \frac{1}{\sqrt{2}}\big(\psi_+(i) - \psi_-(i)\big). \end{aligned} \tag{4.14}$$

In the case of the in-gap states, the resulting ψ_A and ψ_B are *spatially separated.* This accounts, in part, for the fact that the energy splitting of the zero modes, defined as:

$$\delta = 2\langle\psi_A|\mathcal{H}_0|\psi_B\rangle \equiv 2\tilde{t} \tag{4.15}$$

is small. In figure 4.3(c), we plot δ for both the rectangular and the heterojunction nanographene, both with two in-gap states. It is apparent and well known [15] that this quantity decays exponentially with W. In the limit where W is very large (see figure 4.3(c)), δ vanishes, and the energy of the in-gap states goes to $E = 0$, showing that these sublattice polarized states are zero modes [15].

4.3.4 $U \neq 0$

We now consider the effect of interactions and show how it leads to the formation of local moments at the location of the hybridized zero modes [11, 54]. The two energy scales that govern the low energy behavior for the two electrons in the two in-gap

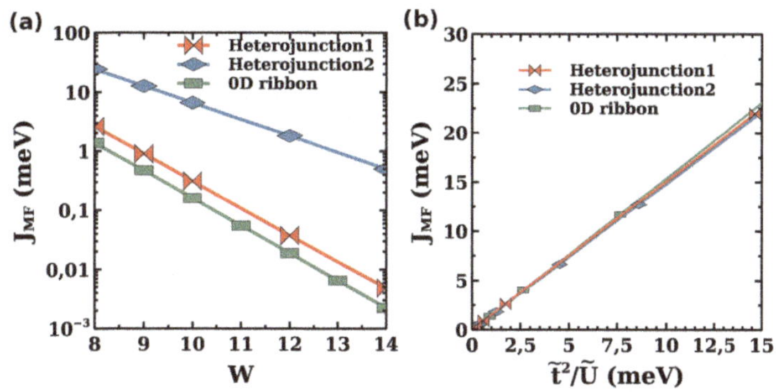

Figure 4.4. Exchange couplings for 0D ribbons and heterojunctions as a function of lateral dimension W and as function of $\frac{\tilde{t}^2}{\tilde{U}}$. In the case of the heterojunctions, we compute independently the coupling mediated by either the wide or the narrow GNR, by adequate choice of the unit cell dimensions.

levels are δ and the energy overhead associated to doubly occupy the sublattice polarized states:

$$\tilde{U} = U\sum_i |\psi_A(i)|^4 = U\sum_i |\psi_B(i)|^4 = U\eta. \tag{4.16}$$

The addition energy is thus given by the product of the atomic Hubbard U and η, is the inverse participation ratio of the zero mode states. Our numerical calculations for the two structures of equation (4.14) yield $\eta = 0.11$ for the zigzag edge states and $\eta = 0.035$ for the interface states. We found that, as opposed to the case of δ, η has a very weak dependence of W. We take $U = |t| = 2.7$ eV. Therefore, the effective Hubbard interaction $\tilde{U}$ is in the range of 270 and 94 meV, for edge and interface states, respectively.

Mean field treatment

We discuss qualitatively the results of a mean field approximation for the Hubbard model for the two nanographenes of equation (4.14). The results are obtained using the collinear mean field treatment (equations (4.6) and (4.7)). For all structures for which $\tilde{U} \gg \delta$ we found broken symmetry solutions with a finite local magnetization, $M(i) = \langle S_z(i)\rangle$ that are mostly located in the region where either ψ_A or ψ_B are non-zero. The results of the magnetization field are shown in the left panels of equation (4.14). The net magnetization per zero mode is close to $S = 1/2$.

Using the mean field approach, we can study the exchange energy as the difference between FM and AF solutions $J_{MF} = E_{FM} - E_{AF}$ as a function of W, for both types of structures. The FM and AF solutions are obtained by suitably forcing the self-consistent iterative procedure to solve the mean field Hamiltonian. We show in figure 4.4 that J_{MF} can be as large as 40 meV can be made small by increasing the distance W between the zero modes. Importantly, as we show in figure 4.4(b), we find that, both for ribbons and heterojunctions, exchange energy scales as

$$J_{MF} \propto \frac{\tilde{t}^2}{\tilde{U}}. \tag{4.17}$$

This scaling provides a strong indication that the mechanism of antiferromagnetic interaction is kinetic exchange [55, 56], which arises naturally for half-filled Hubbard dimers. The fact that local moments are hosted mostly by the in-gap states permits us to build a restricted model where only the in-gap states are considered. This is the topic of the next paragraph.

Effective Hubbard dimer

In order to go beyond the mean field picture and to be able to describe local moments in these nanographenes with a full quantum theory without breaking symmetry, we restrict the Hilbert space to the configurations of two electrons in the two zero modes. To do so, we represent the Hubbard interaction in the one-body basis defined by the states ψ_A and ψ_B. The Hamiltonian so obtained is a two-site Hubbard model with renormalized hopping and on-site energy [11, 54]:

$$\mathcal{H}_{\text{eff}} = \tilde{t}\sum_{\sigma}\left(a_{\sigma}^{\dagger}b_{\sigma} + b_{\sigma}^{\dagger}a_{\sigma}\right) + \tilde{U}(n_{A\uparrow}n_{A\downarrow} + n_{B\uparrow}n_{B\downarrow}) \tag{4.18}$$

where $a_{\sigma}^{\dagger} = \sum_i \psi_A(i)c_{i\sigma}^{\dagger}$ and $b_{\sigma}^{\dagger} = \sum_i \psi_B(i)c_{i\sigma}^{\dagger}$ are the operators that create an electron in the zero modes ψ_A and ψ_B with spin σ, respectively. In turn, $n_{A,\sigma} = a_{\sigma}^{\dagger}a_{\sigma}$ is the number operator for the ψ_A state with spin σ.

Hamiltonian (4.18) is a two-site Hubbard model, where the sites correspond to the zero mode states $\psi_{A,B}$, shown in figures 4.3(b)–(e). For the relevant case of two electrons, the dimension of the Hilbert space is 6 and the ground state is always a singlet, as inferred both from analytical solution [57] or by a straightforward numerical diagonalization [11].

The exact solution permits us to set the language to discuss the emergence of local moments in these structures. For this matter, we can write the wave function of the ground state as:

$$|\Psi_{GS}\rangle = c_2(|2, 0\rangle + |0, 2\rangle) + c_S(|\uparrow, \downarrow\rangle - |\downarrow, \uparrow\rangle) \tag{4.19}$$

where $|2, 0\rangle$ describes a state with two electrons in one site of the Hubbard dimer, and none on the other, whereas $|\sigma_1, \sigma_2\rangle$ describes states with the one electron per site, with spins σ_1, σ_2. In the figure we show how, for $U = 0$, we have $c_2 = c_S = \frac{1}{2}$ so that double occupancy is as likely as individual occupancy. As U is ramped up, the c_2 coefficient is depleted and the c_S coefficient is enhanced, as shown in figure 4.5.

In order to characterize the magnetic behavior of the dimer, we define the spin operators:

$$\begin{aligned} S_z(a) &\equiv \frac{1}{2}\left(a_{\uparrow}^{\dagger}a_{\uparrow} - a_{\downarrow}^{\dagger}a_{\downarrow}\right) \\ S_z(b) &\equiv \frac{1}{2}\left(b_{\uparrow}^{\dagger}b_{\uparrow} - b_{\downarrow}^{\dagger}b_{\downarrow}\right). \end{aligned} \tag{4.20}$$

We can see right away that their expectation values are zero for the ground state, in contrast with the broken symmetry solutions of the mean field theory. We thus

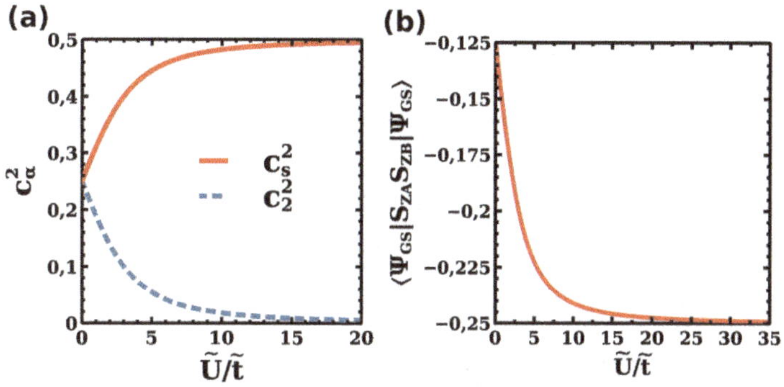

Figure 4.5. Left: evolution of the double occupancy and single occupancy weights in the ground state of the Hubbard dimer at half filling as a function of U/t. Right: evolution of the spin correlation function.

look up at the next moment, the spin correlation function. In particular, we can obtain the following result for the spin correlator for the ground state $\langle\Psi_{GS}|S_z(a)S_z(b)|\Psi_{GS}\rangle = -\frac{c_S^2}{2}$. Thus, for $U = 0$ there is some spin correlation $(-1/8)$. As U is ramped up, the correlation tends to $-1/4$, the value expected for $c_S = 1/\sqrt{2}$ and $c_2 = 0$. In that limit, wave function is identical to the spin singlet of the antiferromagnetic Heisenberg dimer.

In the strong coupling limit, $\tilde{U} \gg \tilde{t}$, it is well known [55, 56] that the four lowest levels in the model of equation (4.18) can be mapped into the Heisenberg Hamiltonian:

$$\mathcal{H}_{\text{Heis}} = J_H \vec{S}_A \cdot \vec{S}_B \tag{4.21}$$

where $J_H \simeq \frac{4\tilde{t}^2}{\tilde{U}}$. The Hamiltonian of equation (4.21) has a ground state singlet $(S = 0)$ as well as an excited state triplet with $S = 1$, separated in energy by $\Delta = E(S = 1) - E(S = 0) = J_H$.

So, the picture that emerges from this model is the following: both structures considered here have two in-gap states each. These in-gap states have a splitting δ that arises from the hybridization of 2 sublattice polarized zero modes. The system is thus modeled with a Hubbard dimer. At half filling, the Hubbard dimer can be effectively mapped into a spin model, when the energy cost of double occupancy of these zero modes, given by equation (4.16) is much larger than the hybridization splitting.

The next question we address is how to up-scale these Hubbard dimers to obtain larger structures. In other words, how to couple more dimers together. There are at least two ways in which we can do this. If we increase the width of the square shape graphene ribbon, making the zigzag edges wider, we shall increase the number of zero modes. Eventually, this leads to the case of 1D channels with ferromagnetic interactions, discussed in the previous section. The other way is discussed in the next section.

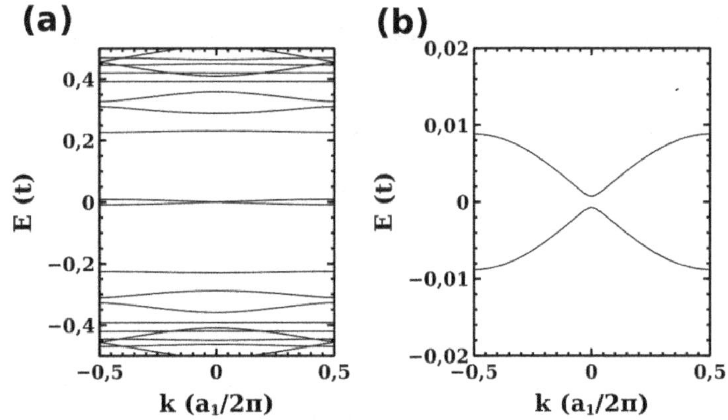

Figure 4.6. (a) Bands of a GNR made of two armchair sections with different widths ($W_t = 12, W_n = 8$). (b) Zoom showing the in-gap narrow bands that arise from the interface zero modes, featuring a dimerization splitting at $k = 0$, coming from the different intra and intercell hybridization of the interface zero modes.

4.4 Dimerized spin chain

We consider a one-dimensional ribbon with armchair edges and alternating section, whose unit cell contains two interface (quasi)-zero modes. The unit cell is described by two length scales, W_t and W_n, that describe the width of the thicker and narrower AGNR. These two length scales control the effective hopping between zero modes. The zero modes thus lead to the formation two bands, inside the gap of the AGNR, shown in figure 4.6. These in-gap bands are effectively described by the t, t' model Hamiltonian. Their bandwidth, governed the effective hybridization of the interface states, can be tuned by changing W, and can easily be much smaller than the band-gap of these ribbons. In figure 4.6, we show a bandwidth of $0.02t$ inside a gap of $0.5t$.

For $t \neq t'$ the two bands have a gap at $k = \pi/L$, where L is the length of the unit cell. The gap closes at $t = t'$. This point separates two insulating phases $t > t'$ and $t < t'$ that are topologically distinct and have a different Zak phase [48]. As a result, only one of them has zero modes in the edges. This happens when the last dimer is affected by the smallest of t and t'

We now consider the effect of interactions in the strong coupling limit $\tilde{U} \gg \tilde{t}, \tilde{t}'$. Using the results of the previous section, and in line with previous work for other GNR structures [51, 58], we consider a spin model [59]

$$\mathcal{H}_{\text{chain}} = \sum_{n=1,N} (J+\delta)\vec{S}_{2n-1}\cdot\vec{S}_{2n} + (J-\delta)\vec{S}_{2n}\cdot\vec{S}_{2n+1}. \tag{4.22}$$

The different exchanges are related to the different hoppings where we can write up:

$$J_n = 4\frac{\tilde{t}_n^2}{\tilde{U}};\quad J_t = 4\frac{\tilde{t}_t^2}{\tilde{U}}\ . \tag{4.23}$$

We now have $J_t - J_n = 2\delta$ and $J_t + J_n = 2J$.

The previous model can be easily solved (4.22) for a chain of $N = 40$ sites (20 dimers) using the density matrix renormalization group [60–66] implemented in the matrix product formalism [67–69] (DMRG). In particular, this method permits us to obtain the expectation values and correlation functions of spin operators the ground state in a computationally efficient manner. The matrix product formalism also allows us to access dynamical quantities of many body systems [70–72]. Specifically, in the following, we discuss the dynamical structure factor defined as

$$A_n(\omega) = \langle \Psi_{GS} | S_n^z \delta(\omega - \mathcal{H} + E_{GS}) S_n^z | \Psi_{GS} \rangle \tag{4.24}$$

The local dynamical structure factor can be qualitatively understood as the quantity giving access to the local density of states of the many-body spin excitations. Using this method, we find that, in bulk, there is a gap for $\delta \neq 0$ (see figure 4.7(b)). The edge structure factor shows gapless states for $\delta < 0$ (see figure 4.7(c)). In figure 4.7(d), we show the map of the spectral function as a function of position and energy, for a fixed value of $\delta = -0.2J$. It is apparent that both edges host zero modes. This phenomenology is similar to the one of the SSH model, and therefore this system can be intuitively understood as a many-body version of a symmetry-protected topological state [73].

The edge topological excitations can be accessed by means of a weak external field. First, it is important to note that, for sufficiently large chains, the dimerized Heisenberg model has a ground state that is four fold degenerate when $\delta < 0$, consisting of a singlet and a triplet state, stemming from the dangling edge excitations. Upon an introduction of a weak external field, the triplet state with $S_z = +1$ becomes the ground state, giving rise to a finite magnetization in the edges. In figure 4.7(e), we show the local expectation value of S_z for the $N = 40$ chain with a small external field of $B_z = 0.01J$.

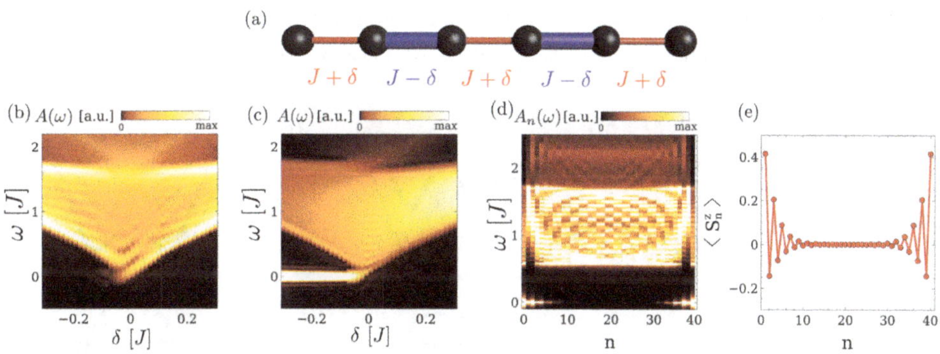

Figure 4.7. (a) Sketch of a dimerized Heisenberg $S = 1/2$ spin chain, having two exchange constants with different strength. Bulk (b) and edge (c) spectral function of the Heisenberg model, showing a gap for $\delta \neq 0$ in the bulk (b), while zero modes for $\delta < 0$ on the edge. (d) The spectral function in every site, showing the edge localization of the zero modes. (e) The local magnetization under a small external field $B_z = 0.01J$, showing the localization of the emergent edge modes.

4.5 Magnetic ribbons competing a superconducting proximity effect

In this section, we discuss a situation where graphene ribbons that host local moments, driven by the exchange interactions discussed above, are in addition exposed to the superconducting proximity effect coming from the substrate. This system permits the study of the competition between magnetism and superconductivity.

We first consider the energy spectrum of the ribbon without magnetic order and with a proximity pairing Δ in the Bogoliubov–de Gennes (BdG) Hamiltonian [40, 74, 75]. As we show in figure 4.8(a), superconducting proximity opens up a gap Δ at the Fermi energy. Time reversal symmetric perturbations can modify the spectrum, but the energy levels cannot be inside the gap.

Things are different when we consider the effect of the magnetic order at the edges. We follow our previous work [75] and we model the exchange by adding a spin and position dependent on-site potential:

$$V_{\text{exch}} = \sum_{i \in \text{edge}} \frac{J(i)}{2}(n_{i\uparrow} - n_{i\downarrow}) \tag{4.25}$$

where $J(i) = 0$ everywhere except at the top edge, for which $J(i) = J$ and the bottom edge, for which $J(i) = \pm J$. We thus consider two different relative orientations of the edge magnetization. The modification of the energy bands and density of states of the superconducting ribbon due to the exchange is shown in figures 4.8 and 4.9, respectively. The most outstanding feature is the emergence of in-gap Yu–Shiba–Rusinov (YSR) [76–78] states. Interestingly, we find an energy dependence of the YSR states linear in J, in line with the one obtained for hydrogenated graphene [75], but different from the standard non-linear dependence of YSR states in normal metals.

The effect of the increasing exchange coupling can be also observed in the Bogoliubov-de-Gennes excitation spectra of the ribbon as shown in figure 4.8. First, in the absence of exchange field, a superconducting gap opens up. As the exchange field increases, the gap starts closing until at critical value the system becomes

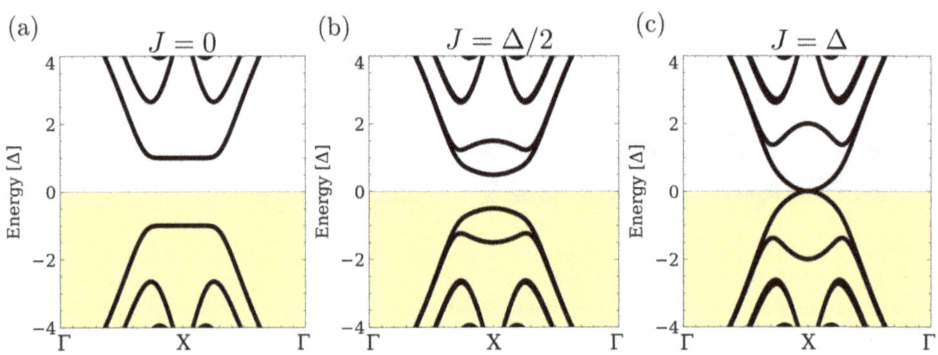

Figure 4.8. (a)–(c) Bogoliubov–de Gennes spectra for a zigzag ribbon with superconducting proximity effect, for different values of the edge exchange J, showing the closing of the gap as J increases. For the widths of the ribbon in 16 atoms, the edges are in the AF configuration and we took $\Delta = 0.2\,t$.

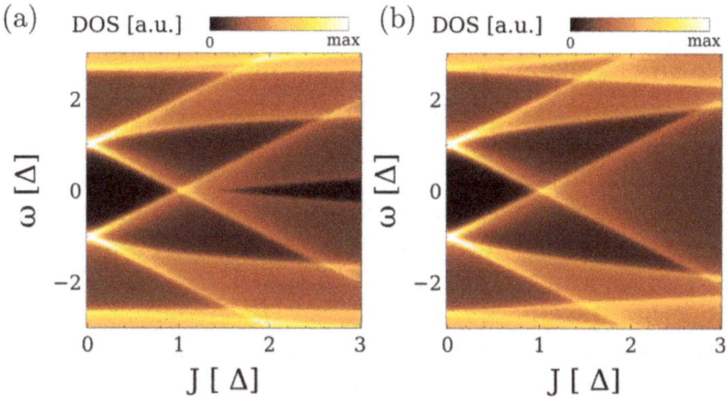

Figure 4.9. Evolution of the DOS for 1D zGNR with superconducting proximity effect as the spin-splitting J induced by ferromagnetic edge order. Left: antiparallel inter-edge alignment. Right: parallel inter-edge alignment. In both instances, in-gap states are induced as J is ramped up. We took a width of 16 atoms and $\Delta = 0.2\,t$.

gapless. This phenomenology is similar to the one found in single magnetic impurities [79], where, as J is increased, a single in-gap excitation approaches the charge neutrality point, giving rise to a parity switching point. In the case of graphene nanoribbons, the behavior is more complex due to the existence of several YSR branches that give rise to a continuum of in-gap excitations [80–84].

4.6 Experimental probes

The experimental study of the non-trivial electronic phases discussed in this chapter can be carried out with state-of-the-art scanning tunnelling microscope (STM) spectroscopy. More specifically, the collective spin excitations of either the magnetically ordered phase, or those of the spin-liquid phase of the dimerized spin chain could be probed with inelastic electron tunneling spectroscopy [6, 85, 86]. Spin excitations in the range of a few meV can be resolved, and the mapping of their intensity profile across the ribbons can help to discriminate from other inelastic excitations in the system, such as phonons.

Given that both the synthesis and the STM probing are required to have the GNR on top of a conducting surface, there will be spin exchange interactions between the local moments at graphene and the conduction electrons at the surface, which we have ignored so far. These Kondo interactions can compete with the exchange interactions discussed so far in several ways. First, if sufficiently strong, the Kondo effect could screen the local moments in graphene. Recent experiments in GNR studied with STM spectroscopy show Kondo peaks [6], which very likely imply the screening of the graphene local moment. Second, additional indirect exchange interactions, mediated by the substrate electrons, can compete with, or perhaps enhance, the graphene mediated interactions. In any event, these Kondo interactions will affect the lineshapes of the inelastic electron tunneling spectra measured with STM [86].

4.7 Conclusions and outlook

We have explored several examples of the emergence of non-trivial quantum phases in graphene nanoribbons. The building blocks for these phases are zero modes that form narrow bands at the Dirac energy. The bandwidth of these bands depends on the aspect ratio of the structures, which thereby provides a control knob. We have mostly focused on the case of neutral GNR that leads to half-full narrow bands, which result in an insulating structure with interesting spin physics but frozen charge dynamics. Departure from half-filling is expected to result in very interesting electronic properties. For instance, doping the 1D ferromagnetic edge is expected to result in domain walls that host fractionalized electrons [27]. Doping the dimerized spin chains might result in superconducting phases. The superconducting proximity effect can be another way to explore the interplay between spin and charge. We have discussed the emergence of in-gap Shiba bands in graphene ribbons. When non-collinear magnetic ground states and/or spin–orbit coupling are considered, the Shiba bands could give rise to topological superconductivity with Majorana end modes [79].

Acknowledgments

J F-R acknowledges financial support from FCT for the P2020-PTDC/FIS-NAN/4662/2014, the P2020-PTDC/FIS-NAN/3668/2014 and the UTAPEXPL/NTec/0046/2017 projects, as well as Generalitat Valenciana funding Prometeo2017/139 and MINECO Spain (Grant No. MAT2016-78625-C2). R O C acknowledges 'Generalitat Valenciana' and 'Fondo Social Europeo' for a PhD fellowship (ACIF/2018/198). J L L acknowledges financial support from the ETH Fellowship program.

References

[1] Wang S, Talirz L, Pignedoli C A, Feng X, Müllen K, Fasel R and Ruffieux P 2016 Giant edge state splitting at atomically precise graphene zigzag edges *Nat. Commun.* **7** 11507

[2] Tao C *et al* 2011 Spatially resolving edge states of chiral graphene nanoribbons *Nat. Phys.* **7** 616

[3] Ruffieux P *et al* 2012 Electronic structure of atomically precise graphene nanoribbons *ACS Nano* **6** 6930–35

[4] Ruffieux P *et al* 2016 On-surface synthesis of graphene nanoribbons with zigzag edge topology *Nature* **531** 489

[5] Jacobse P H, Kimouche A, Gebraad T, Ervasti M M, Thijssen J M, Liljeroth P and Swart I 2017 Electronic components embedded in a single graphene nanoribbon *Nat. Commun.* **8** 119

[6] Li J, Sanz S, Corso M, Choi D J, Peña D, Frederiksen T and Pascual J I 2019 Single spin localization and manipulation in graphene open-shell nanostructures *Nat. Commun.* **10** 200

[7] Yan L and Liljeroth P 2019 Engineered electronic states in atomically precise artificial lattices and graphene nanoribbons arXiv: 1905.03328 (https://ui.adsabs.harvard.edu/abs/2019arXiv190503328Y)

[8] Cao Y *et al* 2018 Correlated insulator behaviour at half-filling in magic-angle graphene superlattices *Nature* **556** 80

[9] Cao Y, Fatemi V, Fang S, Watanabe K, Taniguchi T, Kaxiras E and Jarillo-Herrero P 2018 Unconventional superconductivity in magic-angle graphene superlattices *Nature* **556** 43

[10] Cao T, Zhao F and Louie S G 2017 Topological phases in graphene nanoribbons Junction states, spin centers, and quantum spin chains *Phys. Rev. Lett.* **119** 076401

[11] Ortiz R, García-Martínez N A, Lado J L and Fernández-Rossier J 2018 Electrical spin manipulation in graphene nanostructures *Phys. Rev.* B **97** 195425

[12] Rizzo D J, Veber G, Cao T, Bronner C, Chen T, Zhao F, Rodriguez H, Louie S G, Crommie M F and Fischer F R 2018 Topological band engineering of graphene nanoribbons *Nature* **560** 204–08

Gröning O *et al* 2018 Engineering of robust topological quantum phases in graphene nanoribbons *Nature* **560** 209

[13] Neto A H C, Guinea F, Peres N M R, Novoselov K S and Geim A K 2009 The electronic properties of graphene *Rev. Mod. Phys.* **81** 109

[14] Katsnelson M I 2012 *Graphene: Carbon in Two Dimensions* (Cambridge: Cambridge University Press)

[15] Nakada K, Fujita M, Dresselhaus G and Dresselhaus M S 1996 Edge state in graphene ribbons: nanometer size effect and edge shape dependence *Phys. Rev.* B **54** 17954

[16] Kane C L and Mele E J 2005 Z_2 topological order and the quantum spin Hall effect *Phys. Rev. Lett.* **95** 146802

[17] Sichau J, Prada M, Anlauf T, Lyon T J, Bosnjak B, Tiemann L and Blick R H 2019 Resonance microwave measurements of an intrinsic spin–orbit coupling gap in graphene: a possible indication of a topological state *Phys. Rev. Lett.* **122** 046403

[18] Kou L, Hu F, Yan B, Wehling T, Felser C, Frauenheim T and Chen C 2015 Proximity enhanced quantum spin Hall state in graphene *Carbon* **87** 418–23

[19] Min H, Hill J E, Sinitsyn N A, Sahu B R, Kleinman L and MacDonald A H 2006 Intrinsic and Rashba spin–orbit interactions in graphene sheets *Phys. Rev.* B **74** 165310

[20] Wehling T O, Şaşıoğlu E, Friedrich C, Lichtenstein A I, Katsnelson M I and Blügel S 2011 Strength of effective coulomb interactions in graphene and graphite *Phys. Rev. Lett.* **106** 236805

[21] Fujita M, Wakabayashi K, Nakada K and Kusakabe K 1996 Peculiar localized state at zigzag graphite edge *J. Phys. Soc. Jpn.* **65** 1920–23

[22] Fernández-Rossier J and Palacios J J 2007 Magnetism in graphene nanoislands *Phys. Rev. Lett.* **99** 177204

[23] Fernández-Rossier J 2008 Prediction of hidden multiferroic order in graphene zigzag ribbons *Phys. Rev.* B **77** 075430

[24] Muñoz Rojas F, Fernández-Rossier J and Palacios J J 2009 Giant magnetoresistance in ultrasmall graphene based devices *Phys. Rev. Lett.* **102** 136810

[25] Lado J L and Fernández-Rossier J 2014 Magnetic edge anisotropy in graphenelike honeycomb crystals *Phys. Rev. Lett.* **113** 027203

[26] Lado J L and Fernández-Rossier J 2014 Noncollinear magnetic phases and edge states in graphene quantum Hall bars *Phys. Rev.* B **90** 165429

[27] López-Sancho M P and Brey L 2017 Charged topological solitons in zigzag graphene nanoribbons *2D Mater.* **5** 015026

[28] Jung J and MacDonald A H 2009 Carrier density and magnetism in graphene zigzag nanoribbons *Phys. Rev.* B **79** 235433

[29] Yazyev O V 2010 Emergence of magnetism in graphene materials and nanostructures *Rep. Prog. Phys.* **73** 056501

[30] Soriano D and Fernández-Rossier J 2012 Interplay between sublattice and spin symmetry breaking in graphene *Phys. Rev.* B **85** 195433
[31] Son Y-W, Cohen M L and Louie S G 2006 Energy gaps in graphene nanoribbons *Phys. Rev. Lett.* **97** 216803
[32] Giovannetti G, Khomyakov P A, Brocks G, Kelly P J and van den Brink J 2007 Substrate-induced band gap in graphene on hexagonal boron nitride: *ab initio* density functional calculations *Phys. Rev.* B **76** 073103
[33] Qiao Z, Yang S A, Feng W, Tse W-K, Ding J, Yao Y, Wang J and Niu Q 2010 Quantum anomalous Hall effect in graphene from rashba and exchange effects *Phys. Rev.* B **82** 161414
[34] Yang H-X, Hallal A, Terrade D, Waintal X, Roche S and Chshiev M 2013 Proximity effects induced in graphene by magnetic insulators: first-principles calculations on spin filtering and exchange-splitting gaps *Phys. Rev. Lett.* **110** 046603
[35] Hallal A, Ibrahim F, Yang H, Roche S and Chshiev M 2017 Tailoring magnetic insulator proximity effects in graphene: first-principles calculations *2D Mater.* **4** 025074
[36] Cardoso C, Soriano D, Garca-Martnez N A and Fernández-Rossier J 2018 van der Waals spin valves *Phys. Rev. Lett.* **121** 067701
[37] Jackiw R and Rebbi C 1976 Solitons with fermion number 1/2 *Phys. Rev.* D **13** 3398–409
[38] Lado J L and Fernández-Rossier J 2015 Quantum anomalous Hall effect in graphene coupled to skyrmions *Phys. Rev.* B **92** 115433
[39] Leutenantsmeyer J C, Kaverzin A A, Wojtaszek M and van Wees B J 2016 Proximity induced room temperature ferromagnetism in graphene probed with spin currents *2D Mater.* **4** 014001
[40] Beenakker C W J 2006 Specular Andreev reflection in graphene *Phys. Rev. Lett.* **97** 067007
[41] Yazyev O V and Helm L 2007 Defect-induced magnetism in graphene *Phys. Rev.* B **75** 125408
[42] Palacios J J, Fernández-Rossier J and Brey L 2008 Vacancy-induced magnetism in graphene and graphene ribbons *Phys. Rev.* B **77** 195428
[43] González-Herrero H, Gómez-Rodrguez J M, Mallet P, Moaied M, Palacios J, Salgado C, Ugeda M M, Veuillen J Y, Yndurain F and Brihuega I 2016 Atomic-scale control of graphene magnetism by using hydrogen atoms *Science* **352** 437–41
[44] García-Martínez N A, Lado J L, Jacob D and Fernández-Rossier J 2017 Anomalous magnetism in hydrogenated graphene *Phys. Rev.* B **96** 024403
[45] Santos E J G, Ayuela A and Sánchez-Portal D 2012 Universal magnetic properties of sp^3-type defects in covalently functionalized graphene *New J. Phys.* **14** 043022
[46] Sutherland B 1986 Localization of electronic wave functions due to local topology *Phys. Rev.* B **34** 5208–11
[47] Soriano D, Muñoz Rojas F, Fernández-Rossier J and Palacios J J 2010 Hydrogenated graphene nanoribbons for spintronics *Phys. Rev.* B **81** 165409
[48] Delplace P, Ullmo D and Montambaux G 2011 Zak phase and the existence of edge states in graphene *Phys. Rev.* B **84** 195452
[49] Kim W Y and Kim K S 2008 Prediction of very large values of magnetoresistance in a graphene nanoribbon device *Nat. Nanotechnol.* **3** 408
[50] Yazyev O V and Katsnelson M I 2008 Magnetic correlations at graphene edges: basis for novel spintronics devices *Phys. Rev. Lett.* **100** 047209
[51] Cornelie K and Wessel S 2017 Quantum phase transitions in effective spin-ladder models for graphene zigzag nanoribbons *Phys. Rev.* B **96** 165114

[52] Shi Z and Affleck I 2017 Effect of long-range interaction on graphene edge magnetism *Phys. Rev.* B **95** 195420

[53] Raczkowski M and Assaad F F 2017 Interplay between the edge-state magnetism and long-range Coulomb interaction in zigzag graphene nanoribbons: quantum Monte Carlo study *Phys. Rev.* B **96** 115155

[54] Golor M, Koop C, Lang T C, Wessel S and Schmidt M J 2013 Magnetic correlations in short and narrow graphene armchair nanoribbons *Phys. Rev. Lett.* **111** 085504

[55] Anderson P W 1959 New approach to the theory of superexchange interactions *Phys. Rev.* **115** 2–13

[56] Moriya T 1960 Anisotropic superexchange interaction and weak ferromagnetism *Phys. Rev.* **120** 91

[57] Jafari J A 2008 Introduction to hubbard model and exact diagonalization *Iran. J. Phys. Res.* **8** 116

[58] Golor M, Wessel S and Schmidt M J 2014 Quantum nature of edge magnetism in graphene *Phys. Rev. Lett.* **112** 046601

[59] Chitra R, Pati S, Krishnamurthy H R, Sen D and Ramasesha S 1995 Density-matrix renormalization-group studies of the spin-1/2 Heisenberg system with dimerization and frustration *Phys. Rev.* B **52** 6581–87

[60] White S R 1992 Density matrix formulation for quantum renormalization groups *Phys. Rev. Lett.* **69** 2863–66

[61] Schollwöck U 2005 The density-matrix renormalization group *Rev. Mod. Phys.* **77** 259–315

[62] Jeckelmann E 2002 Dynamical density-matrix renormalization-group method *Phys. Rev.* B **66** 045114

[63] Schollwöck U 2011 The density-matrix renormalization group in the age of matrix product states *Ann. Phys.* **326** 96–192

[64] Stoudenmire E M and White S R 2017 Sliced basis density matrix renormalization group for electronic structure *Phys. Rev. Lett.* **119** 046401

[65] White S R 2005 Density matrix renormalization group algorithms with a single center site *Phys. Rev.* B **72** 180403

[66] Manmana S R, Stoudenmire E M, Hazzard K R A, Rey A M and Gorshkov A V 2013 Topological phases in ultracold polar-molecule quantum magnets *Phys. Rev.* B **87** 081106

[67] Verstraete F, Murg V and Cirac J I 2008 Matrix product states, projected entangled pair states, and variational renormalization group methods for quantum spin systems *Adv. Phys.* **57** 143–224

[68] Stoudenmire E M and White S R 2013 Real-space parallel density matrix renormalization group *Phys. Rev.* B **87** 155137

[69] Itensor http://itensor.org/

[70] Weiße A, Wellein G, Alvermann A and Fehske H 2006 The kernel polynomial method *Rev. Mod. Phys.* **78** 275–306

[71] Wolf F A, McCulloch I P, Parcollet O and Schollwöck U 2014 Chebyshev matrix product state impurity solver for dynamical mean-field theory *Phys. Rev.* B **90** 115124

[72] Lado J L and Zilberberg O 2019 Topological spin excitations in Harper-Heisenberg spin chains arXiv: 1906.07090

[73] Su W P, Schrieffer J R and Heeger A J 1979 Solitons in polyacetylene *Phys. Rev. Lett.* **42** 1698–701

[74] San-Jose P, Lado J L, Aguado R, Guinea F and Fernández-Rossier J 2015 Majorana zero modes in graphene *Phys. Rev.* X **5** 041042

[75] Lado J L and Fernández-Rossier J 2016 Unconventional Yu–Shiba–Rusinov states in hydrogenated graphene *2D Mater.* **3** 025001

[76] Yu L 1965 Bound state in superconductors with paramagnetic impurities *Acta Phys. Sin.* **21** 75–91

[77] Hiroyuki S 1968 Classical spins in superconductors *Prog. Theor. Phys.* **40** 435–51

[78] Rusinov A I 1969 Theory of gapless superconductivity in alloys containing paramagnetic impurities *Sov. Phys. JETP* **29** 1101–06

[79] Balatsky A V, Vekhter I and Zhu J-X 2006 Impurity-induced states in conventional and unconventional superconductors *Rev. Mod. Phys.* **78** 373–433

[80] Nadj-Perge S, Drozdov I K, Li J, Chen H, Jeon S, Seo J, MacDonald A H, Bernevig B A and Yazdani A 2014 Observation of Majorana fermions in ferromagnetic atomic chains on a superconductor *Science* **346** 602–7

[81] Röntynen J and Ojanen T 2015 Topological superconductivity and high chern numbers in 2D ferromagnetic Shiba lattices *Phys. Rev. Lett.* **114** 236803

[82] Ménard G C *et al* 2015 Coherent long-range magnetic bound states in a superconductor *Nat. Phys.* **11** 1013–16

[83] Ruby M, Peng Y, von Oppen F, Heinrich B W and Franke K J 2016 Orbital picture of Yu-Shiba-Rusinov multiplets *Phys. Rev. Lett.* **117** 186801

[84] Kezilebieke S, Dvorak M, Ojanen T and Liljeroth P 2018 Coupled Yu–Shiba–Rusinov states in molecular dimers on $NbSe_2$ *Nano Lett.* **18** 2311–23

[85] Hirjibehedin C F, Lutz C P and Heinrich A J 2006 Spin coupling in engineered atomic structures *Science* **312** 1021–24

[86] Fernández-Rossier J 2009 Theory of single-spin inelastic tunneling spectroscopy *Phys. Rev. Lett.* **102** 256802

[87] Zhang Y-H, Kahle S, Herden T, Stroh C, Mayor M, Schlickum U, Ternes M, Wahl P and Kern K 2013 Temperature and magnetic field dependence of a Kondo system in the weak coupling regime *Nat. Commun.* **4** 2110

Chapter 5

Transport in graphene nanoribbon-based systems

Leonor Chico, Jhon W González, Marta Pelc and Hernán Santos

5.1 Introduction

Despite the excellent transport properties of graphene [1], the lack of an electronic gap and the impossibility to open it in its two-dimensional form leads to size reduction as an alternative for its application in nanoelectronics [2]. Quantum size effects may open a gap in graphene by rolling graphene into nanotubes or cutting it into ribbons, which stand out as prospective components for future nanoelectronics [3]. Another path to open a gap in graphene is to consider Bernal-stacked bilayer graphene (BLG), and apply an electric field [4, 5]. The maximum gap obtained is around 0.3 eV, providing an extra tool to modify the conductance of the system. However, for electronic applications, larger gaps are desirable, leaving ultranarrow nanoribbons as the most promising candidates for quasi-one-dimensional carbon circuitry. In fact, bottom-up dehalogenation techniques demonstrated that few Å-wide ribbons with perfect edges can be synthesized with atomic precision [6]. This approach has also been recently applied to the fabrication of nanoribbon hetero-junctions [7] and molecular devices with narrow ribbons as electrical contacts [8].

Graphene nanoribbons (GNRs) can be metals or semiconductors depending on their edge geometry, if electron interactions are neglected [9]. First-principles calculations predict that all ribbons are gapped, albeit the origin of the gap depends on the type of edge termination [10]. For zigzag-terminated ribbons, localized edge states dominate their transport properties; they have been put forward as a way to employ these systems in spintronics and magnetic applications. Being quasi-one-dimensional systems, pristine nanoribbons present quantized conductance. Therefore, defects and contacts play a crucial role in the transport behavior of nanoribbon-based devices.

In this chapter, we review the transport properties of quasi-one-dimensional systems based on graphene nanoribbons. We focus on those with metallic behavior,

 © IOP Publishing Ltd 2020

more relevant for transport applications. We also analyze the role of nanoribbons as valley filters for carbon nanotubes, as well as their possible application for spintronic devices.

5.2 Description of the systems

Since our main interest is the low-energy transport properties of graphene-based systems, we employ either a one-orbital tight-binding (TB) description or the continuum Dirac model to describe the electronic behavior of these structures. The Landauer–Kubo approach is followed to compute the quantum conductances in the low-bias zero temperature limit.

We first describe unzipped carbon nanotubes, which can be seen as a hybrid system composed of a carbon nanotube (CNT) seamlessly joined to a nanoribbon, or a sequence of these junctions, as shown in figure 5.1. If the nanotube is of the armchair type, it yields a zigzag graphene nanoribbon (ZGNR) upon unzipping. The band structure of an armchair nanotube has two inequivalent valleys separated in momentum space, like graphene. Zigzag nanoribbons are metallic within the one-electron TB approximation; inclusion of electron repulsions via a Hubbard term opens a small gap, but the localized character of the edge bands is preserved.

We also consider nanoribbons connected to a bilayer flake. Such a system can be built either by placing a graphene flake on top of a nanoribbon or overlapping two semi-infinite graphene strips; in both cases, the structure can be modeled as composed by two monolayer ribbons, which act as contacts, connected to a bilayer graphene flake. These two contact arrangements are depicted in figure 5.2(a). Depending on the contact geometry and the type of stacking, transport responses can be quite different.

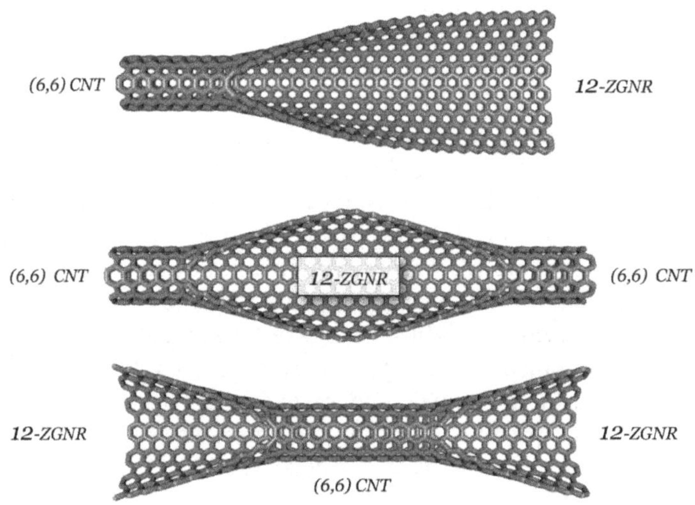

Figure 5.1. Schematic geometry of three possible unzipped nanotubes. (Top) A (6,6) CNT/12-ZGNR junction; center: (6,6) CNT/12-ZGNR/(6,6) CNT quantum dot. (Bottom) A 12-ZGNR/(6,6) CNT/12-ZGNR quantum dot.

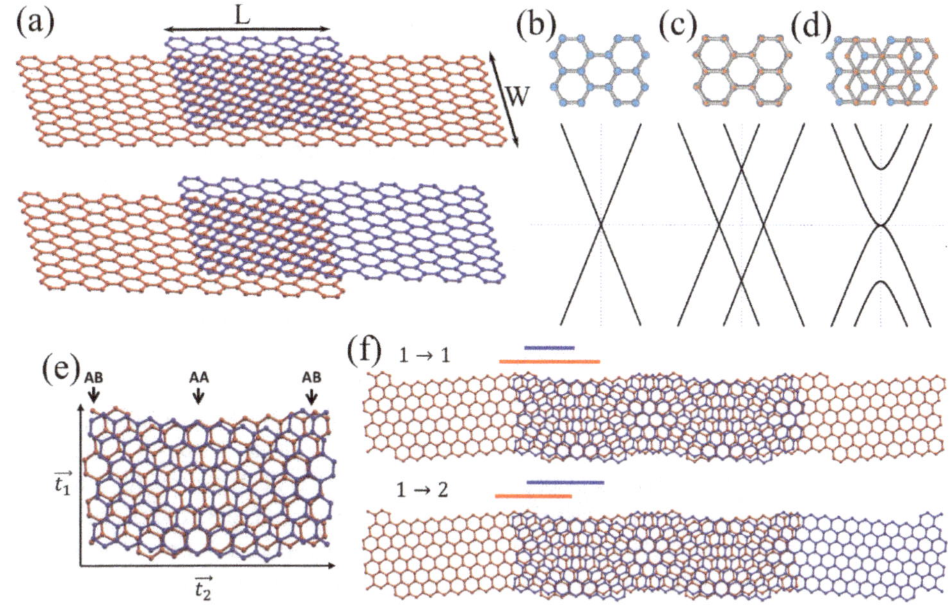

Figure 5.2. (a) Bilayer graphene flakes connected to two semi-infinite nanoribbons. (Upper plot) Overlaid flake on top of a ribbon. (Lower plot) Bilayer flake formed by two overlapping nanoribbons. The width and length of the bilayer region are W and L, respectively. (b)–(d) Atomic structure geometries and band dispersion relations around the Dirac point for several armchair-terminated nanoribbons. The longitudinal axes of the ribbons are in the horizontal x direction. (b) Monolayer armchair GNR; (c) bilayer GNR with AA stacking; (d) bilayer GNR with AB-α stacking. For this energy range, the dispersion relations for metallic armchair ribbons (b)–(d) are independent of the ribbon width. (e) Unit cell of the (10, 1)[(4, 3)] twisted bilayer nanoribbon. (f) Twisted bilayer analogues of the structures depicted in (a). Reprinted with permission from [11], Copyright (2010) by the American Physical Society. Reprinted with permission from [12], Copyright (2015) by the American Physical Society.

With respect to the stacking, bilayer graphene appears in three main flavors: if the two atoms in the unit cell of each graphene layer are equally labeled as A and B, in the *AA* stacking both atoms of one layer, A1 and B1, are located exactly on top of the equivalent atoms of the other layer, say A2 and B2, as depicted in figure 5.2(c). Its band structure resembles that of monolayer graphene, consisting of two massless Dirac cones split by the interlayer interaction, as schematically shown therein. Because of sublattice symmetry, no gaps are opened upon applying an electric field. In Bernal or AB stacking, the two graphene layers are arranged in such a way that the A1 sublattice is exactly below sublattice B2. This leaves atom B1 (A2) at the center of a hexagon of the other layer 2 (1), as shown in figure 5.2(d). Due to the breaking of sublattice geometry, its bands are massive with approximately a parabolic dispersion near the neutrality point. Additionally, a gap can be opened by an external electric field.

The minimum distance between non-equivalent A and B atoms in the same layer is as in graphite or graphene, $a_{CC} = 1.42$ Å. The separation between layers depends on the type of stacking. For Bernal or AB stacking the distance is $c_{\mathrm{AB}} = 3.35$ Å [13], while in AA stacking the distance is slightly higher, $c_{AA} = 3.55$ Å [14]. These experimental distances are in accordance with predictions made using *ab-initio* calculations for BLG [15–17] and bilayer GNRs including van der Waals interactions [18]. In any case, the distance between atoms belonging to different layers in both stackings is much larger than the separation between atoms in the same layer, a_{CC}. For AA or AB bilayer GNRs, we focus on metallic armchair-terminated systems, such as those depicted in figure 5.2(c) and (d). In Bernal-stacked bilayer ribbons, two possible edge arrangements are possible, depending on whether the edge atoms are aligned on the two ribbons or not. We concentrate here on the most symmetrical stacking, called AB-α [11, 19], which allows for the analytical treatment of their electronic properties.

Besides these two simple stackings, the two graphene layers can be rotated with respect to each other, creating the so-called twisted bilayer graphene (TBG) [20–25]. Depending on the rotation angle, the structures can be commensurate (i.e. periodic) or not; in any case, for small rotation angles (more clearly seen below 10°), a moiré pattern emerges due to the misalignment between the two layers. Incommensurate bilayers are harder to model theoretically; due to the phase mismatch between layers, they are assumed to behave as composed of independent single layers [26]. However, recent experiments indicate that coupling is not negligible in incommensurate TBG [27]; further work on these intriguing systems is needed.

Commensurate TBG presents a superperiodic structure; band structure calculations can be performed to elucidate its properties. Different electronic regimes have been identified [24, 28, 29]. For large rotation angles, the system behaves as monolayer graphene, as if the two sheets of graphene were uncoupled. With decreasing rotation angle (below 10°), a gradual decrease of the Fermi velocity can be observed. Below 2° flat bands develop, and around 1° the Fermi velocity drops to zero [30–32]. This is known as the magic angle, and there are theoretical predictions for the occurrence of more magic angles below 1° [24].

From the experimental viewpoint, TBG was found in few-layer samples, but initially the rotation angle could not be tuned at will. This situation has changed recently; new techniques, based on breaking a single graphene flake, and rotating one of the layers in a controlled manner, have opened the possibility to study particular angles on demand [33]. Naturally, interest has concentrated on magic angles, for which correlated behavior [33, 34] and unconventional superconductivity have been recently reported [35]. Thus, the study of twisted graphene systems and the interplay of flat and edge bands will surely be an active topic in graphene research.

5.3 Modeling graphene nanoribbon systems

Near the charge neutrality point, the electronic properties of graphene and graphene-like materials are well described by p_z orbitals. In the energy region of interest for the transport properties, graphene bands are linear, so they can be also modeled by a continuum Dirac Hamiltonian.

5.3.1 TB approach

For monolayer nanoribbons, we use a one-orbital π-band Hamiltonian considering only first-neighbor interactions, with a hopping parameter γ_0 around 3 eV. In this chapter, values from 2.66 to 3.16 eV are employed; the differences are minimal, amounting to a trivial scaling of the bands. The same values are valid for carbon nanotubes.

If edge effects are taken into account, as in the case of zigzag edges in unzipped nanotubes, a Hubbard term is added to the Hamiltonian, which reads

$$H = -\gamma_0 \sum_{i,j,\sigma} c_{i,\sigma}^{\dagger} c_{j,\sigma} + U \sum_{i,\sigma} n_{i,\sigma} < n_{i,-\sigma} > , \tag{5.1}$$

where $c_{i,\sigma}^{\dagger}(c_{j,\sigma})$ is a creation (annihilation) operator at atom $i(j)$ of an electron with spin σ, $n_{i,\sigma}$ is the corresponding occupation number operator, and U is the onsite Coulomb interaction term. For the ribbons studied here, U ranging from 2 to 3 eV correctly describes the low-energy bands of ZGNRs.

For bilayer graphene structures, only in-plane nearest-neighbor hoppings are included, as in the monolayer cases. The interlayer coupling depends on the type of stacking. For AA- or AB-stacked bilayers, the interlayer interaction is modeled with a single hopping parameter, $\gamma_1 = 0.1\,\gamma_0$, only connecting atoms directly on top of each other. For the sake of simplicity, we have discarded other Hamiltonian terms, such as trigonal warping, as their effects become relevant only when we move away from charge neutrality point.

The TB Hamiltonian for the AB bilayer is

$$H^{\mathrm{AB}} = -\gamma_0 \sum_{<i,j>,m} \left(a_{m,i}^{+} b_{m,j} + h.\ c.\right) - \gamma_1 \sum_{i} \left(a_{1,i}^{+} b_{2,i} + h.\ c.\right), \tag{5.2}$$

where $a_{m,i}(b_{m,i})$ annihilates an electron on sublattice A(B), in layer $m = 1, 2$, at site i. $\langle i, j \rangle$ indicates that the sum is over nearest-neighbors. The second term in equation (5.2) corresponds to the layer-layer interaction in the AB-stacking. For the bilayer with AA-stacking, the Hamiltonian reads

$$H^{\mathrm{AA}} = -\gamma_0 \sum_{<i,j>,m} \left(a_{m,i}^{+} b_{m,j} + h.\ c.\right) - \gamma_1 \sum_{i} \left(a_{1,i}^{+} a_{2,i} + b_{1,i}^{+} b_{2,i} + h.\ c.\right). \tag{5.3}$$

For twisted bilayer systems, the Hamiltonian is given by $H = H_1 + H_2 + H_{12}$, where 1, 2 are the layer indices. The layer Hamiltonians H_1 and H_2 only contain nearest-neighbor interactions; since the TBG hopping parameters are obtained by fits to the *ab-initio* results, we take the in-plane hopping $\gamma_0 = 3.16$ eV. The interlayer Hamiltonian H_{12} is given by

$$H_{12} = \sum_{i,j} \left(-\gamma_1^{-\beta(r_{ij}-d)} c_i^{\dagger} c_j + h.\ c.\right). \tag{5.4}$$

Here, $\gamma_1 = 0.12\ \gamma_0$ is the interlayer hopping, r_{ij} is the difference between in-plane coordinates of atoms i and j that belong to different layers, $d = 3.35$ Å is the distance between layers, and $\beta = 3$ is a parameter fitted to the density functional theory calculations [22, 36]. The cutoff for the interlayer interaction is set to $6a_{CC}$.

Within the TB approximation, the conductance and local densities of states (LDOS) of non-periodic systems are calculated using a Green function (GF) matching approach [37, 38]. The system is divided into three parts: a central region (the conductor) connected to the right and left leads. The Hamiltonian is therefore $H = H_C + H_R + H_L + h_{LC} + h_{RC}$, where H_C, H_L, and H_R are the Hamiltonians of the conductor, left and right leads, respectively, and h_{LC}, h_{RC} are the hoppings from the left L and right R leads to the conductor C. The GF of the latter is $\mathcal{G}_C(E) = (E - H_C - \Sigma_L - \Sigma_R)^{-1}$, where $\Sigma_\ell = h_{\ell C} g_\ell h_{\ell C}^\dagger$ is the self-energy due to lead $\ell = L, R$, and $g_\ell = (E - H_\ell)^{-1}$ is the GF of lead ℓ.

In a zero bias approximation, the Landauer conductance G is given by $G = \frac{2e^2}{h} T(E_F)$. T is the transmission function that can be obtained from the GFs of the central part and the leads:

$$T(E) = \mathrm{Tr}\big(\Gamma_L(E)\mathcal{G}_C(E)\Gamma_R(E)\mathcal{G}_C^\dagger(E)\big), \tag{5.5}$$

where $\Gamma_\ell = i(\Sigma_\ell - \Sigma_\ell^\dagger)$ describes the coupling of C to lead ℓ. For a single junction, the left and right electrodes are connected directly, so $\mathcal{G}_C$ in equation (5.5) is just the GF of the interface between the left and right lead, $\mathcal{G}_{LR}$.

Finally, the local density of states is given by

$$\mathrm{LDOS}(E) = -\frac{1}{\pi}\mathrm{Im}(\mathrm{Tr}(\mathcal{G}_C(E))); \tag{5.6}$$

here the trace is taken over all the nodes of the conductor. In certain cases, the atom-resolved LDOS (without summing over the nodes) is plotted, in order to analyze the spatial distribution of the state in the system.

5.3.2 Continuum Dirac approximation

The low-energy properties of monolayer and bilayer graphene systems are well described with the $\mathbf{k}\cdot\mathbf{p}$ approximation [39]. For monolayer graphene, a linear, Dirac-like Hamiltonian is obtained. Bilayer Hamiltonians can be written using the same approach, with similar characteristics [5, 40].

For AA-stacked bilayer graphene, the low-energy effective Hamiltonian is

$$H_{\mathrm{AA}} = \begin{pmatrix} 0 & v_F\pi^\dagger & \gamma_1 & 0 \\ v_F\pi & 0 & 0 & \gamma_1 \\ \gamma_1 & 0 & 0 & v_F\pi^\dagger \\ 0 & \gamma_1 & v_F\pi & 0 \end{pmatrix}, \tag{5.7}$$

where $\pi = k_x + ik_y = ke^{i\theta_k}$, $\theta_k = \tan^{-1}(k_x/k_y)$, and $\mathbf{k} = (k_x, k_y)$ is the momentum relative to the Dirac point. The Fermi velocity v_F is the slope of the Dirac linear bands, which is related to the TB hopping parameter by $v_F = \frac{\sqrt{3}}{2}\gamma_0 a$. The Hamiltonian acts on a four-component spinor $\left(\phi_A^{(1)}, \phi_B^{(1)}, \phi_A^{(2)}, \phi_B^{(2)}\right)$. The eigenfunctions of this Hamiltonian are bonding and antibonding combinations of the isolated graphene sheet solutions,

$$\varepsilon_{s,\pm}^{AA} = sv_F k \pm \gamma_1 , \quad \psi_{s,\pm}^{AA} = \begin{pmatrix} 1 \\ se^{i\theta_k} \\ \pm 1 \\ \pm se^{i\theta_k} \end{pmatrix} e^{i\mathbf{k}\cdot\mathbf{r}}, \tag{5.8}$$

with $s = \pm 1$.

The continuum Hamiltonian for AB-stacked graphene is [5]

$$H_{AB} = \begin{pmatrix} 0 & v_F\pi^\dagger & 0 & \gamma_1 \\ v_F\pi & 0 & 0 & 0 \\ 0 & 0 & 0 & v_F\pi^\dagger \\ \gamma_1 & 0 & v_F\pi & 0 \end{pmatrix}, \tag{5.9}$$

with eigenvalues

$$\varepsilon_{s,\pm}^{AB} = \frac{s}{2}\left(\gamma_1 \pm \sqrt{4v_F^2 k^2 + \gamma_1^2}\right), \quad s = \pm 1. \tag{5.10}$$

The wavefunction corresponding to an eigenvalue E can be found analogously; it has a more complicated expression that will not be used here. For details, see [11].

In order to describe monolayer and bilayer graphene nanoribbons (BGNRs), the appropriate boundary conditions should be imposed to the wavefunctions. This yields the allowed wavevectors, eigenvalues and eigenfunctions of the GNRs. By matching the corresponding GNR wavefunctions, the transmission through the system can be obtained.

5.4 Zigzag graphene nanoribbons as valley filters

A novel use of ZGNRs as valley filters was proposed in 2009, without resorting to external potentials [41]. The idea stemmed from the electronic properties of nanostructures composed of carbon nanotubes partially unzipped into graphene nanoribbons. As mentioned above, these systems can be seen as junctions of CNTs and ribbons. They have been experimentally synthesized as a byproduct of techniques aimed at obtaining GNRs from longitudinal unzipping of CNTs, either by chemical attack [42], plasma etching [43], or lithium intercalation followed by exfoliation [44]. Partially open CNTs have been recently used to fabricate cascaded logic circuits [45], showing its potential applicability in nanoelectronics.

If the carbon nanotube is of the armchair type, it has two inequivalent valleys contributing to the conductance. Opening it, a zigzag graphene nanoribbon is obtained, which acts as a valley filter for the parent CNT. In order to illustrate the valley filtering behavior, we choose a specific junction made of a (6,6) CNT and a

12-ZGNR, such as the one depicted at the top of figure 5.1. Armchair tubes are labeled as (n, n), where $2n$ is the number of carbon atoms around the circumference, and ZGNRs by the number N of zigzag chains along its width, N-ZGNR.

Here, we present our results within the TB approximation [41]; the electronic conductance is obtained employing equation (5.5). Analytical results have been derived by other authors within the TB approach [46], but only for the non-interacting case. *Ab-initio* calculations have allowed us to describe wider energy ranges [47, 48], but the physical behavior of these systems are best understood with simpler models.

The conductance of a (6,6) CNT/12-ZGNR junction assuming non-interacting electrons is presented in figure 5.3(a). It shows that around the Fermi energy E_F (from −0.85 eV to +0.85 eV, first plateau), the ZGNR acts as a perfect valley filter for the armchair tube. Assuming left to right conduction, the ZGNR only has the K' channel open to transport around E_F, as shown in the inset. It is completely

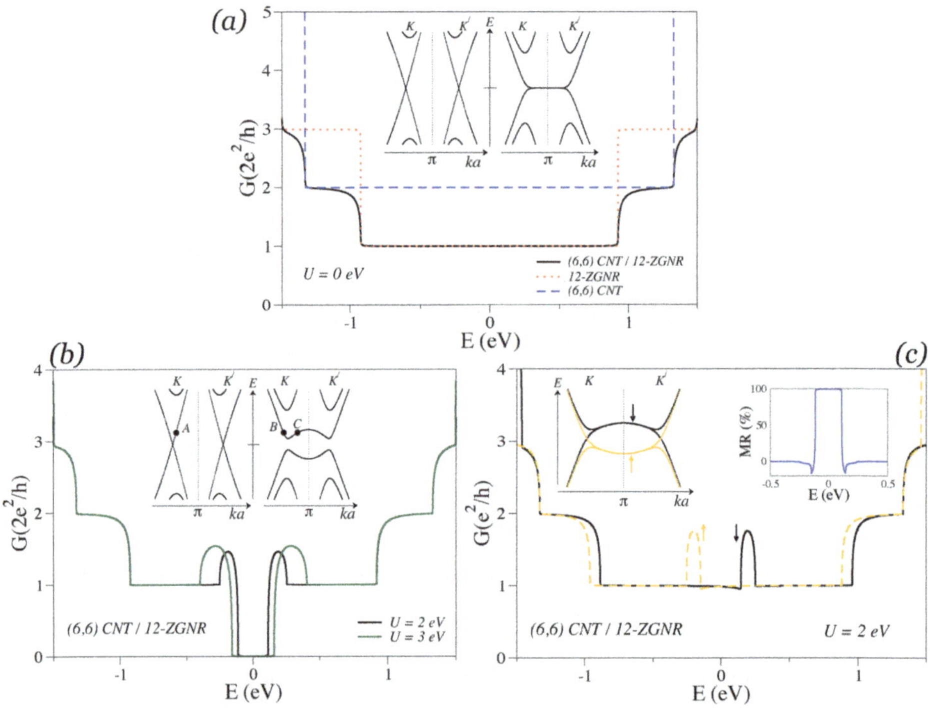

Figure 5.3. (a) Black solid line: conductance of a (6,6) CNT/12-ZGNR system. For comparison, the conductances of the perfect (6,6) CNT (12-ZGNR) in dashed blue (red dotted) lines. All conductances are computed assuming non-interacting electrons. The inset shows a schematic plot of the low-energy band structures of the CNT and the ZGNR with their two valleys. (b) Conductance of a (6,6) CNT/12-ZGNR system including Coulomb interactions with two U values. The inset shows the CNT and ZGNR with $U = 3$ eV band structures. (c) Conductance of a (6,6) CNT/12-ZGNR system in the FM configuration. The inset shows the spin-split bands of the ribbon, black (orange) for spin parallel (antiparallel) to the magnetic field. The inset shows the energy dependence of the magnetoresistance. Reprinted with permission from [41], Copyright (2009) by the American Physical Society.

transparent to states from the K' valley of the CNT, blocking the incident carriers from K. For energies above 0.85 eV (or below −0.85 eV), a new conductance channel in the GNRs permits conduction from both valleys, so the valley filtering behavior is lost. The valley filtering effect is robust for larger systems (i.e. larger diameter CNTs, as (18,18) CNT/36-ZGNR), or when the GNR is narrower than the one derived from the parent tube, as in a (6,6) CNT/10-ZGNR junction, provided that the energy range only allows carriers from one valley in the ribbon.

Inclusion of electronic interaction changes the scenario close to E_F because of the magnetic properties induced in the ZGNR. Magnetic moments appear at the zigzag edges, with a ferromagnetic (FM) coupling along the same edge and an antiferromagnetic (AFM) coupling between the two edges. Thus, the non-interacting edge flat bands at E_F acquire a certain dispersion if the Coulomb interaction is included, and a gap opens. This gap increases with U, so that a transport gap develops near E_F, as shown in figure 5.3(b). Above the gap, there is a region of enhanced conductance with respect to the noninteracting case. This is due to the dispersion of the bands close to E_F, which permits transmission from K to K'. This is illustrated in the inset of figure 5.3(b): state A in the K valley of the CNT can be transmitted into state C of the ZGNR which belongs to the same valley, contributing to a conductance enhancement over $2e^2/h$. State C is an edge state, located on both edges with opposite spin polarization. Therefore, the extra transmission takes place at the edges, with spin-polarized conductances. Above this interval (below for hole conduction), the valley filtering properties are recovered, as can be seen in figure 5.3(b).

Applying a magnetic field $\boldsymbol{B}$ perpendicular to the GNR, the situation can change dramatically. The Zeeman coupling competes with the AFM ordering of the magnetic moments at the edges, and if $\boldsymbol{B}$ is sufficiently large, the moments at the edges flip to FM-aligned and the gap closes. The FM band structure close to E_F is plotted in the inset of figure 5.3(c). In this arrangement, the ZGNR is ferromagnetic and metallic. The bands are split and cross at zero energy due to electron–hole symmetry. This generates a non-zero value of the conductance near E_F, for which the AFM solution is insulating. Therefore, the magnetoresistance, defined as $MR(\%) = \frac{R_{\mathrm{FM}} - R_{\mathrm{AFM}}}{R_{\mathrm{FM}} + R_{\mathrm{AFM}}} \cdot 100$, where $R_{\mathrm{FM(AFM)}}$ is the ferromagnetic (antiferromagnetic) resistance, reaches a maximum value in the gap energy range. The results of the MR that are indicated in the right inset of figure 5.3(c) show a magnetoresistance close to 100% near the Fermi energy. Therefore, besides the valley filter, this system is by itself a spin filter featuring 100% magnetoresistance.

The properties of other nanostructures, such as those depicted in the middle and lower sections of figure 5.1, have been also studied [41], showing that the transparency and the valley and spin-filter properties also hold, being robust in these carbon systems.

5.5 AA- and AB-stacked bilayer nanoribbon flakes

The simplest geometry to explore the transport properties of a bilayer graphene flake is to employ monolayer graphene nanoribbon contacts, as depicted in figure 5.2(a).

Two straightforward arrangements are considered: either an overlaid flake on top of a monolayer ribbon, as in the upper part of the figure, or two overlapping GNRs, depicted at the bottom part. These are the nanoribbon analogues of the telescoping nanotubes or the CNT shuttles [49]. An advantage of the overlapping ribbon setup is that the bilayer length can be tuned by displacing the top ribbon. Although it may seem too idealized, this type of junction has been recently built and characterized in transport experiments of graphene nanoribbon break junctions [50], corroborating our theoretical predictions of oscillation periods larger than the graphene lattice constant, consistent with a dependence on the bilayer flake length.

Since the transmission through bilayer flakes depends on many variables besides their length, we restrict ourselves to the simplest geometries; moreover, we choose them to have armchair edges. In this way, an analytical derivation of the transmission within the Dirac model is possible, providing a unique insight into the properties of these devices. Of course, geometrical variations, such as different widths for the ribbons or flakes, could be also studied, but for the sake of simplicity, we stick to same-size, symmetrically stacked armchair GNRs (AGNR) (see figures 5.2(c) and (d)). We start by analytical results derived from the Dirac-like Hamiltonians, which provide an invaluable tool to elucidate the transport behavior of these systems.

5.5.1 Continuum model: transmission from wavefunction matching

We consider metallic armchair nanoribbons, i.e. those with widths $N = 3p + 2$, where $p = 0, 1, 2, \ldots$. The number of dimers across the width of the ribbon is indicated by N, as is customarily done. For monolayer ribbons, the two Dirac points merge in one (see figure 5.2(b)). As indicated therein, electrons incide in the x direction. The lowest subbands correspond to $k_y = 0$; with these assumptions, boundary conditions are imposed at the beginning ($x = 0$) and at the end ($x = L$) of the bilayer flake. Either the wavefunctions of the monolayer are connected to one of the layers of the flake, and continuity of the wavefunctions is imposed, or the layer terminates, so the vanishing of the wavefunctions at the ends is required.

AA stacking

In the $1 \to 1$ (bottom-bottom) configuration, the wavefunction should be continuous in the bottom layer, i.e. $\phi_{\mathrm{A}}^{(1)}(x)$ continuous at $x = 0$ and $\phi_{\mathrm{B}}^{(1)}(x)$ continuous at $x = L$, modeling the junction between the monolayer ribbon and the bottom of the bilayer flake. For the top layer, the wavefunction vanishes at the edges of the flake:

$$\phi_{\mathrm{A}}^{(2)}(x = 0) = \phi_{\mathrm{B}}^{(2)}(x = L) = 0. \tag{5.11}$$

From these boundary conditions, we obtain the transmission

$$T_{\mathrm{AA}}^{1\to 1} = 1 - \frac{\sin^4 \dfrac{\gamma_1 L}{v_F}}{1 + 2\cos\dfrac{2EL}{v_F}\cos^2\dfrac{\gamma_1 L}{v_F} + \cos^4\dfrac{\gamma_1 L}{v_F}}. \tag{5.12}$$

In the $1 \to 2$ configuration, the bottom wavefunction $\phi_A^{(1)}(x)$ and the top wavefunction $\phi_B^{(2)}(x)$ should be continuous at $x = 0$ and $x = L$, respectively. In addition, the hard-wall condition at the flake terminations should be satisfied:

$$\phi_A^{(2)}(x=0) = \phi_B^{(1)}(x=L) = 0. \tag{5.13}$$

The above boundary conditions yield the transmission

$$T_{AA}^{1\to 2} = 1 - \frac{\cos^4 \dfrac{\gamma_1 L}{v_F}}{2\left(1 - \cos\dfrac{2EL}{v_F}\right)\sin^2\dfrac{\gamma_1 L}{v_F} + \cos^4\dfrac{\gamma_1 L}{v_F}}. \tag{5.14}$$

From both expressions, equations (5.12) and (5.14), it can be seen that the conductance, $G = (2e^2/\hbar)T$, changes periodically as a function of the energy and length of the bilayer region. For fixed L, the transmission is a periodic function of the energy with quasilocalized states in the top part of the bilayer flake at energies given by $\frac{\pi v_F}{L}(n + \frac{1}{2})$, with $n = 0, 1, 2\ldots$. In the bottom-top configuration, the wavevectors of the quasilocalized states of the bilayer flake are shifted by $-\frac{\pi}{2L}$. If the energy is fixed, the conductance varies periodically with the length of the bilayer region L. There is a period, $\pi v_F/E$, related to the energy of the incident carrier; other periods, harmonics of that imposed by the interlayer hopping $\pi v_F/\gamma_1$, also appear.

In the limit $E \to 0$, the transmission of the overlapping ribbons goes to zero,

$$T_{AA}^{1\to 2}(E=0) = 0, \tag{5.15}$$

and the bottom-bottom case has a very simple expression:

$$T_{AA}^{1\to 1}(E=0) = 1 - \frac{4\sin^4\dfrac{\gamma_1 L}{v_F}}{3 + \cos^2\dfrac{2\gamma_1 L}{v_F}}. \tag{5.16}$$

In these last two expressions, the phase opposition between the $1 \to 1$ and $1 \to 2$ transmissions is most evident. For the bottom-top configuration, there is always a minimum zero transmission at $E = 0$ irrespective of the system size. In the complementary $1 \to 1$ configuration, the transmission is a maximum, but its value depends on the length of the flake and the interlayer coupling. We will see later that this complementarity also holds for other stackings.

AB stacking

Recall that in this stacking only the atoms A of layer 1 and the atoms ***B*** of layer 2 are directly connected by the interlayer hopping (see equation (5.9)). The energies of the carriers in the central bilayer flake are given by equation (5.10). For an incident carrier with $|E| > \gamma_1$ and wavevector k_x there are two reflected and two transmitted eigenfunctions with wavevector $\pm k_{1(2)} = \pm\sqrt{k_x(k_x \pm \gamma_1/v_F)}$ in the bilayer region.

However, for incident wavefunctions with $|E| < \gamma_1$, there are only one reflected and one transmitted central wavefunctions with $\pm k_1 = \pm\sqrt{k_x(k_x + \gamma_1/v_F)}$. There are additionally an exponentially growing and a decaying state with decay constants $\kappa = \pm\sqrt{k_x(\gamma_1/v_F - k_x)}$. Thus, the conductance of the system depends on the energy of the carrier being larger or smaller than the interlayer hopping γ_1. For $|E| > \gamma_1$, there are two channels in the central region and the interference between these channels produces antiresonances (zero transmission), whereas for $|E| < \gamma_1$ only one electronic channel is present in the central region. In this case, Fabry–Pérot interference yielding conductance oscillations can occur within the central bilayer region.

By employing the same boundary conditions (continuity and hard-wall at the ends of the flakes, equations (5.11) and (5.13)), analytical, but cumbersome, impractical expressions, can be obtained for the conductance in the AB-α stacking. Even in the low and high energy limit, the transmissions $T_{\rm AB}^{1\to1}$ and $T_{\rm AB}^{1\to2}$ have rather complicated expressions. Therefore, we choose to describe the main features of the transmissions, and illustrate them with some selected numerical results. Details for the analytical formulae can be found in [11]. At $E = 0$, the two configurations present a perfect complementary behavior: $T_{\rm AB}^{1\to1}(E = 0) = 1$, whereas $T_{\rm AB}^{1\to2}(E = 0) = 0$. For $|E| < \gamma_1$, since there is only one conducting channel in the bilayer, resonances with the bilayer length L appear in both contact geometries. For large L, they occur if $L = (n + \frac{1}{4})\frac{\pi}{k_1}$ (n integer) in the $1 \to 1$ configuration; they are shifted with respect to the overlapping $1 \to 2$ case, which presents them at $L = (n + \frac{1}{2})\frac{\pi}{k_1}$. If $|E| > \gamma_1$, the transmission shows antiresonances associated with destructive interferences of the two electronic paths in the bilayer region for both configurations. The lower harmonic in the AB stacking, $\frac{k_1 - k_2}{2}L \sim \frac{\gamma_1 L}{2v_F}$, is half the basic harmonic in the AA stacking, and this reflects the fact that in the AB stacking only half of the atoms have direct interlayer tunneling. For $E \gg \gamma_1$, periodicities related to the energy of the incident electron and flake size L can be detected, such as $2EL/v_F$, and others related to harmonics of the interlayer hopping, with an overall approximate complementarity.

5.5.2 Electronic conductances within the TB and Dirac approaches

Figure 5.4 shows the conductance of a $N = 17$ AGNR system as a function of the bilayer flake length for two energies, above and below the interlayer hopping, and for both configurations. The length of the bilayer flake L is given in terms of the translational unit cell, and the continuum results are directly compared to the discrete TB calculations. As commented before, the electron transmission across the bilayer flake presents oscillations as a function of the length flake L, being a spatial period larger than the lattice constant, regardless of the system configuration ($1 \to 1$ or $1 \to 2$) and the stacking (AA or AB) [11, 51, 52], as has been experimentally verified [50].

The complementarity of the conductances in the $1 \to 1$ and $1 \to 2$ is quite evident in this figure, most notably at $E = 0$, as stated above. If the energy is fixed, the electron transmission clearly oscillates with the length of the bilayer flake L. For the AA-stacked bilayer, the transmission for the $1 \to 2$ and $1 \to 1$ configurations are

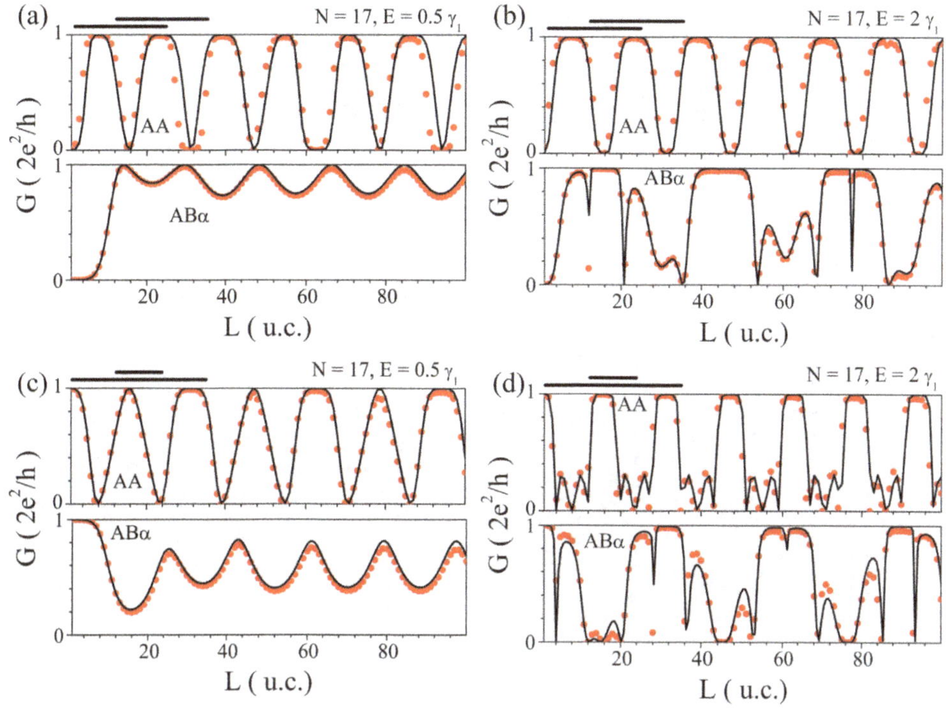

Figure 5.4. Conductance as a function of the bilayer region length L for an AGNR of width $N = 17$ with AA and AB-α stackings, for an incident carrier with energy $E = \gamma_1/2$ in (a), (c) and $E = 2\gamma_1$ in (b), (d). (a), (b) The $1 \to 2$ configuration and (c), (d) the $1 \to 1$ configuration. Red circles: tight-binding results. Black solid lines: Dirac model calculations. Reprinted with permission from [11], Copyright (2010) by the American Physical Society.

clearly in counterphase; see for example the maxima in figure 5.4(a) (top panel) with the minima in figure 5.4(c) (top panel). It can also be appreciated for the AB-α stacking around $E = 0$. A similar behavior can be seen when comparing figure 5.4(b) with figure 5.4(d), although for this energy, $E > \gamma_1$, the conductance presents in general more structure.

Figure 5.5 shows the conductance as a function of the flake length L and energy E for both stackings and lead configurations. In the AA-stacked bilayer graphene flakes, additionally to the oscillation of the conductance with the flake size L, there is a periodicity related the energy E of the carrier, as discussed at the sight of the analytical formulae. For fixed L, the transmission is a periodic function of the energy, and the complementary behavior of the conductance for both configuration is most apparent for this plot.

The behavior of the electronic transmission of AB-stacked graphene flakes presents two different regimes defined by interlayer hopping γ_1. For an incident carrier with $|E| > \gamma_1$ there are always two reflected and two transmitted eigenfunctions in the bilayer region, as always occurs for AA-stacked bilayer. In this regime, the interference between the conducting channels produces antiresonances, and the conductance as a function of

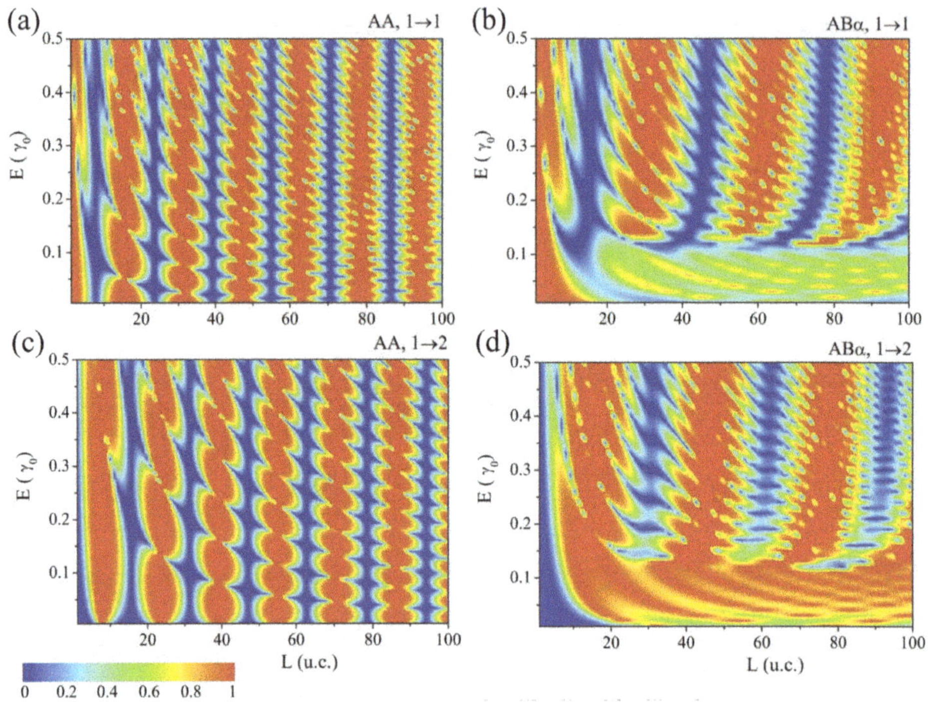

Figure 5.5. Color plot of the conductance as a function of the bilayer region length L and energy of the carrier E for a metallic AGNR with AA and AB-α stackings, as obtained from the continuum Dirac model. (a), (b) The $1 \rightarrow 1$ configuration and (c), (d) the $1 \rightarrow 2$ configuration. Reprinted with permission from [11], Copyright (2010) by the American Physical Society.

the length is characterized by the oscillation between zero and the maximum transmission. However, for incident wavefunctions with $|E| < \gamma_1$, there are only one reflected and one transmitted wavefunctions at AB-stacked flakes. The conductance of this regime is dominated by Fabry–Pérot interference processes. The fingerprint of this regime is the existence of oscillations in the electron transmission that do not reach zero. This behavior does not occur in AA-stacked systems.

When the number of available conducting channels for both stackings is the same, the electronic transmission exhibits similar properties. For instance, for a carrier with an incident energy of $|E| = 2\gamma_1$ the electron transmission of both stackings presents antiresonances with zero conductance. As mentioned before, difference in the period between AA- and AB-α stacking bilayer systems is related to the effective interlayer coupling, larger for the AA case, yielding half the period in L than the AB-α stacked flakes.

Finally, we would like to mention that the application of an external gate provides an additional degree of freedom to tune the transport properties of these systems; González *et al* [51] provides further results along this line.

5.6 Twisted bilayer graphene nanoribbons

Twisted bilayer graphene nanoribbon devices such as those presented in the previous section have unique features that distinguish them from the symmetrically stacked systems. The interplay between moiré regions with different local stackings and edges is distinctive of these nanostructures and worth exploring. TBG flake edges contain AA- and AB-stacked regions where low-energy localized states appear, and govern the transport properties near E_F. This localization at the AA- and AB-stacked edges can take place far from the flake-lead interfaces. As a consequence, in this energy range the transport does not depend on the leads configuration. The complementary behavior seen in the AA- and AB-α bilayer GNR flakes only occurs far from the Fermi level, where states are spread over the whole flake.

We construct the unit cell of the twisted bilayer nanoribbon following [12, 53], by cutting it from a twisted bilayer graphene sheet. The rotation axis is set in one atom of the B sublattice, and the top layer is rotated so that an atom with coordinates $\vec{r} = m\vec{a}_1 + n\vec{a}_2$ falls into an equivalent site $\vec{t}_1 = n\vec{a} + m\vec{a}_2$, where $\vec{a}_1 = \frac{a}{2}(\sqrt{3}, -1)$ and $\vec{a}_2 = \frac{a}{2}(\sqrt{3}, 1)$ are the graphene lattice vectors. The relative rotation angle (RRA) is defined by $\cos\theta = (n^2 + 4nm + m^2)/2(n^2 + nm + m^2)$. We focus on systems with $n = m + 1$. In order to have ribbons with predominantly zigzag-terminated edges, we choose the rectangular unit cell spanned by the vectors $\vec{t}_1 = n\vec{a} + m\vec{a}_2$ and $\vec{t}_2 = (2m + n, n - m) = (3m + 1, 1)$[1]. We label the nanoribbons by giving the coordinates of vector $\vec{t}_2$ and the corresponding two-dimensional moiré unit cell coordinates in brackets, that is $(3m + 1, 1)[(n, m)]$. A minimal nanoribbon width is chosen here, that is, $|\vec{t}_1|$. Unless stated otherwise, the length of the flake is equal to two unit cells of the twisted bilayer ribbon. This allows for distinguishing stacking-related localization from that due to the finite size of the flake. The edges of the ribbons are designed to be minimal (with a minimum number of edge atoms with coordination number 2 [54, 55]). The results given below correspond to (10, 1) [(4, 3)] bilayer flakes, where RRA is $\theta = 9.43°$ and the number of atoms in two unit cells is $2N = 592$.

The left panels of figures 5.6(a) and (b) show the bilayer-averaged LDOS in both $1 \to 1$ and $1 \to 2$ configurations, and the right panels present the corresponding conductances. Depending on the energy ranges, we observe a different transport behavior. Near E_F, for $|E| < 0.3$ eV, LDOS peaks are correlated to conductance maxima. In this energy range, the infinite twisted ribbon has at most one conducting channel, as depicted in figure 5.6(c). For higher energies, $|E| > 0.3$ eV, when the bilayer conductor has two available channels, antiresonances take place [56]. Thus, there are two different transport regimes, either resonant transport through a single channel, or destructive interference between two possible channels. Both behaviors were also observed in the more symmetric stackings, AA and AB-α, discussed the

[1] If the edge termination is not important, a better choice of the unit cell with rhombic shape and half the size can be found [53]

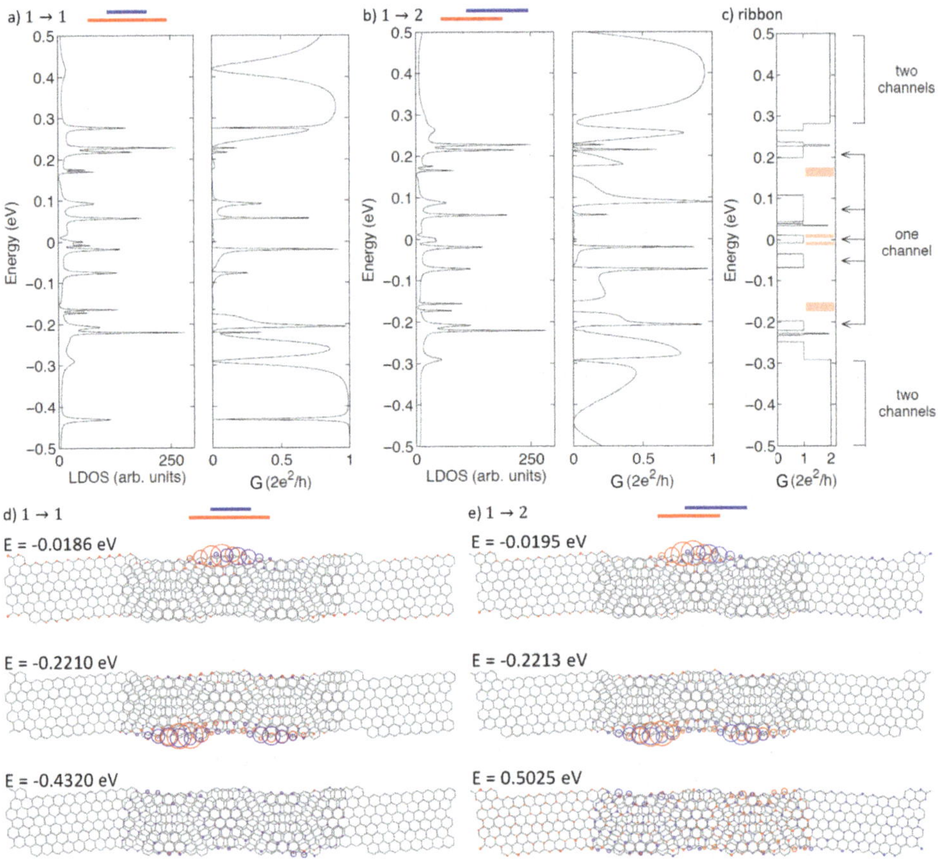

Figure 5.6. LDOS and conductance plots for a (10, 1)[(4, 3)] TBG flake: (a) 1 → 1 configuration; (b) 1 → 2 configuration. (c) Conductance of a (10, 1)[(4, 3)] TBG infinite nanoribbon. Pink rectangles mark the energy ranges of zero-conductance for the leads, made of single-layer ribbons (10, 1). The flake length is equal to two unit cells ($2|\vec{t}_2|$), approximately 50 Å, and its width $|\vec{t}_1|$ is about 15 Å. LDOS for three states of the TBG nanoribbon flakes with different state localizations: (d) 1 → 1: $E = -0.0186$ eV edge state with AB-stacking; AA-stacked edge state at $E = -0.2210$ eV; and flake state at $E = -0.4320$ eV; (e) 1 → 2: $E = -0.0195$ eV, edge state with AB stacking; $E = -0.2213$ eV, edge state with AA stacking; and $E = 0.5025$ eV, flake state. The radii of the circles are proportional to the LDOS on each atom. For the lowest state of (e), LDOS values are multiplied by 20. Color indicates localization on the bottom (red) and top (blue) layer. Reprinted with permission from [12], Copyright (2015) by the American Physical Society.

previous subsection [11, 51]. In particular, the AA stacking always showed the two-channel interference, whereas for AB-α stacking, it depended on the carrier energy.

Indeed, the electronic and transport characteristics of the leads and central conductor are key to understand the overall behavior of the leads-flake system. The conductance gaps of the monolayer leads, marked with pink rectangles in (c) correspond to zero-conductance values in the whole system; right of figures 5.6(a) and (b). However, the gaps in the bulk bilayer conductor, such as from 0.1 to 0.2 eV or from −0.2 to −0.06 eV (figure 5.6(c)) do not necessarily prevent the charge flow.

Broad conductance peaks and shoulders can be seen in the right panels of figures 5.6(a) and (b) due to tunneling through the flake, here only two unit cells long.

Importantly, there is a distinct difference in the conductance of TBG systems and those with more symmetric stackings studied in the previous section. In the twisted flakes, for energies $|E| < 0.3$ eV, the localized states giving rise to transmission resonances are due to edge states. Therefore, the behavior of the two lead configurations is quite similar, and in fact they even have close energies. In figures 5.6(d) and (e), the atom-resolved LDOS of several flake states are presented, along with their corresponding energies. The top states in both panels, with energies close to E_F, are clear examples of AB-edge localized states, while the middle states are mostly localized at AA-stacking edges. This is in agreement with the typical edge localization in infinite twisted ribbons [53]. The edge character of the LDOS keeps them away from the connections to the leads, making them independent of the particular lead configurations.

However, farther from the Fermi level, for $|E| > 0.3$ eV, we observe different behaviors in the two lead configurations. These states are spread all over the flake region, as depicted in figures 5.6(d) and (e). Thus, they are sensitive to the dissimilar boundary conditions for the bilayer flake. In both geometries, the LDOS maxima correspond to antiresonances in the conductance. Notwithstanding, their character is different. In the case of the $1 \rightarrow 1$ system, the state is mostly localized in the top layer. The opposite $1 \rightarrow 2$ configuration has its states localized in both layers.

Finally, the transport dependence on the flake length, N, also shows different behavior depending on the energy range. Figure 5.7 shows the conductance at an energy $E = 0.35$ eV, with two propagating channels in the bilayer flake for the two lead configurations. Large period oscillations with antiresonances are apparent, and the $1 \rightarrow 1$ and $1 \rightarrow 2$ configurations reveal a clear complementary behavior, just like in the case AA- and AB-stacking flakes. If the energy falls in the conductance gap of the bilayer flake, the conductance presents an exponential decay with N, and no complementarity shows up, as in the energy range governed by edge-localized states.

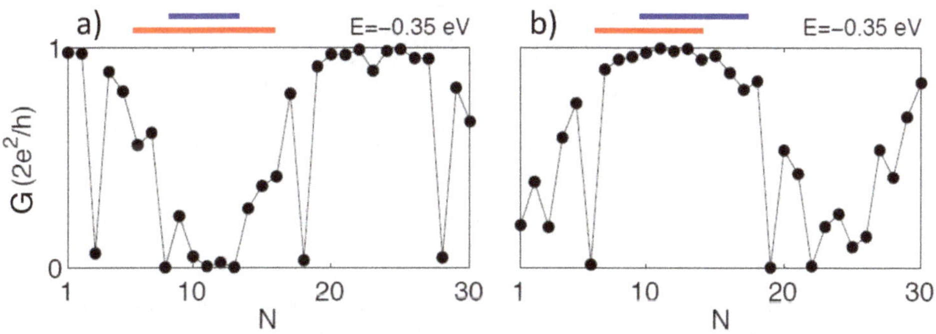

Figure 5.7. Conductance as a function of flake length (N) calculated for both system configurations for the energy value $E = -0.35$ eV. Reprinted with permission from [12], Copyright (2015) by the American Physical Society.

In summary, twisted bilayer graphene flakes show many similarities to the AA- and AB-stacked ribbons described in the previous section, if the energy of the carriers is above the edge-moiré localized states. However, for low-energy transport, twisted bilayer ribbons show a unique behavior, not seen in more symmetric stackings.

5.7 Spin-polarized transport in graphene nanoribbons

At the nanoscale, the production of strong polarized currents is being explored extensively in different systems, and GNRs are no exception. The use of magnetic electrodes is the most direct way to achieve a spin-polarized current in any device [56], but magnetic fields are difficult to harness at nanometer sizes. Spin manipulation via electric fields is an alternative path explored in the last few years via spin-orbit coupling (SOC), to the extent that this emerging field has been dubbed spin-orbitronics [57]. Graphene, material that presents a wealth of groundbreaking applications, was seminally proposed in this spintronics scenario as a possible quantum spin Hall insulator [58]. However, carbon has a very small SOC, and only via curvature, external fields or by proximity effects can it be enhanced [59–61].

Notwithstanding, GNRs provide an excellent playground to explore the production of spin-polarized currents by Rashba spin–orbit interactions [62]. The variety of spatial symmetries present in ideal graphene nanoribbons permits the exploration of this phenomenon with a single system. The proposed device is composed by a GNR as shown in figure 5.8(a). The central region of the ribbon is subjected to an external electric field, switching on the Rashba SOC therein. So within this region, a Rashba term [62, 63] is added to the nearest-neighbor, one-orbital Hamiltonian of the ribbon, given by equation (5.1) with $U = 0$.

Depending on the symmetry of the system, an electric-field-induced Rashba coupling can produce a spin-polarized current without the need for breaking time reversal symmetry. It is therefore the most relevant to elucidate which symmetries in planar systems allow for such spin polarization of the current.

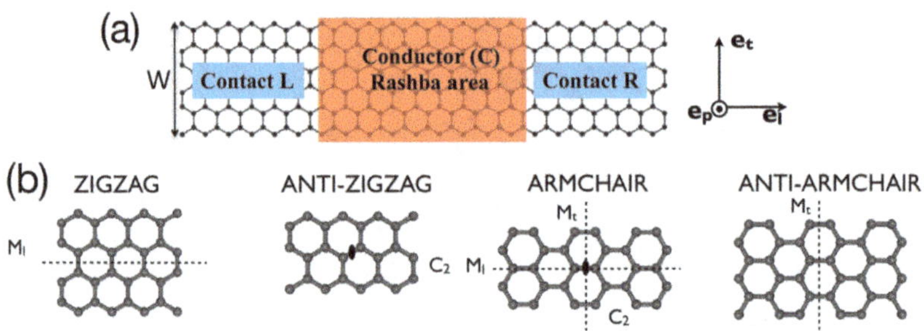

Figure 5.8. (a) Schematic view of the device geometry, a nanoribbon in which the central part, the conductor (C) highlighted in red, experiences Rashba SOC induced by an external electric field. The spin directions considered here are shown in blue and labeled e_l, e_t and e_p. (b) Geometries of the central conductor flakes subject to SOC, with their corresponding spatial symmetries. Reprinted with permission from [62], Copyright (2015) by the American Physical Society.

The spin-resolved conductance is computed in the Kubo approach by using the Green function formalism, as explained in subsection 5.3.1, but including the spin. The conductances have an explicit spin dependence and an explicit indication of the current direction: thus, in $G^{LR}_{\sigma_i\sigma'_i}(E)$ LR means the direction of the current. Note that i is the direction into which the spin σ is projected, i.e. transversal (t), longitudinal (l), and perpendicular (p) to the plane of the device (see figure 5.8(a)). The spin polarization of the current in the i-direction is defined as $P_i(E) = G^{LR}_{\sigma_i\sigma_i} - G^{LR}_{\sigma_i\bar{\sigma}_i} + G^{LR}_{\bar{\sigma}_i\sigma_i} - G^{LR}_{\bar{\sigma}_i\bar{\sigma}_i}$, where $\bar{\sigma}_i$ stands for $-\sigma_i$. In the small-bias limit, $P_i(E)$ is proportional to the i-component of the spin current.

The existence of spin-polarized currents can be explained in terms of the lack of certain symmetries of the system, mainly spatial and time-reversal. This can be applied not only to GNRs, taken here as archetypal examples, but in general to any planar devices.

Figure 5.9(a) shows the conductance relationships given by symmetry relations, summarized for the four studied GNR shapes. As an example, ZGNR flakes present spatial longitudinal mirror symmetry, M_l, which is responsible for the relation $G^{LR}_{\sigma\sigma'} = G^{LR}_{\bar{\sigma}\bar{\sigma}'}$ for longitudinal and perpendicular spin directions, e_l and e_p, while for the transversal direction, e_t, gives $G^{LR}_{\sigma\sigma'} = G^{LR}_{\sigma\sigma'}$. This means that only spin-polarized currents in ZGNR can be obtained if the spin is projected into the transversal direction, due to $G^{LR}_{\uparrow\uparrow} \neq G^{LR}_{\downarrow\downarrow}$ and $G^{LR}_{\uparrow\downarrow} \neq G^{LR}_{\downarrow\uparrow}$. Other cases can be similarly followed in the table-summary of figure 5.9(a). An anti-ZGNR presents a C_2 rotational spatial symmetry around the perpendicular ribbon axis, an anti-AGNR has transversal mirror symmetry and finally, the AGNR has all the spatial symmetries, M_l, C_2 and M_t. The combination of these symmetries with time-reversal may yield relations

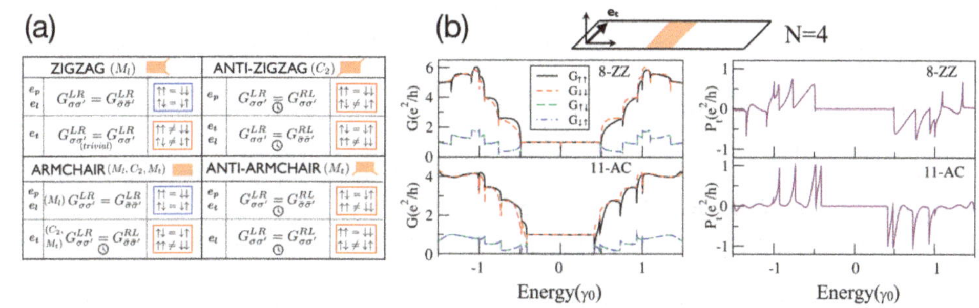

Figure 5.9. (a) Summary of the results presented in [62]. For each flake (and type of symmetry), the first column lists the spin projection directions; the second column indicates the corresponding spin-resolved conductance relation along with the spatial symmetries employed for its derivation (time reversal symmetry is shown as a clock when it is invoked); the third column indicates whether the spin-resolved conductances are equal or not due to symmetry, and when spin-polarized current are feasible, those results are enclosed in a red box (either because spin-conserved polarizations are different ($\uparrow\uparrow \neq \downarrow\downarrow$) or spin-flip polarizations are not equal ($\uparrow\downarrow \neq \downarrow\uparrow$)). A blue box encloses the corresponding equalities when no spin polarization is possible for symmetry reasons. (b) Spin-resolved conductances (left panel) and spin polarization of the current (right panel) for two types of GNR, 11-armchair (labeled AC in the figure) and 8-zigzag (ZZ). The length of the flake is $N = 4$, and the spin projection direction is schematically shown in the top of the panels. Reprinted with permission from [62], Copyright (2015) by the American Physical Society.

between spin-flip and the spin-conserved conductances, from which the possibility of spin-polarized currents can be elucidated.

Numerical results corroborate the previous analysis; for more computational instances, see [62]. Figure 5.9(b) shows the conductance and polarization P_t projected in the transversal direction of two types of GNRs. Spin-conserved conductances, $G_{\uparrow\uparrow}$ and $G_{\downarrow\downarrow}$, reach one conductance quantum close to E_F for both cases, indicating its metallic behavior. It is noteworthy that beyond the first plateau the two conductances are unequal, giving rise to a spin-polarized current. Spin-flip conductances, $G_{\uparrow\downarrow}$ and $G_{\downarrow\uparrow}$, look like identical, although there is a small difference for the ZGNR case.

The spin polarization in the other spin directions, P_l and P_p (not shown here) are not so large as P^t. This is due to the nature of the Rashba spin–orbit interaction: the Rashba term takes the form $H_R \propto (\sigma \times \mathbf{k}) \cdot \mathbf{E}$. Thus, a maximum contribution can be expected when the directions of electric field **E**, the current direction **k** and the spin are orthogonal, as it happens when the spin is pointing in the transversal direction.

These symmetry considerations have been also applied to other complex systems as CNTs [64], for which the relevant symmetry operations are more general than those studied here. Additionally, the presence of topological defects or adsorbed atoms provides a mechanism to enhance the spin-polarized currents [65].

References

[1] Novoselov K S, Geim A K, Mozorov S V, Jiang D, Zhang Y, Dubonos S V, Gregorieva I V and Firsov A A 2004 Electric field effect in atomically thin carbon films *Science* **306** 666

[2] Castro Neto A H, Guinea F, Peres N M R, Novoselov K S and Geim A K 2009 The electronic properties of graphene *Rev. Mod. Phys.* **81** 109–62

[3] Baringhaus J *et al* 2014 Exceptional ballistic transport in epitaxial graphene nanoribbons *Nature* **506** 349

[4] Ohta T, Bostwick A, Seyller T, Horn K and Eli Rotenberg K 2006 Controlling the electronic structure of bilayer graphene *Science* **313** 951–4

[5] McCann E and Fal'ko V I 2006 Landau-level degeneracy and quantum Hall effect in a graphite bilayer *Phys. Rev. Lett.* **96** 086805

[6] Cai J *et al* 2010 Atomically precise bottom-up fabrication of graphene nanoribbons *Nature* **466** 470

[7] Cai J *et al* 2014 Graphene nanoribbon heterojunctions *Nat. Nanotechnol.* **9** 896–909

[8] Li J, Merino-Dez N, Carbonell-Sanromà E, Vilas-Varela M, de Oteyza D G, Peña D, Corso M and Ignacio Pascual J 2018 Survival of spin state in magnetic porphyrins contacted by graphene nanoribbons *Sci. Adv.* **4** eaaq0582

[9] Nakada K, Fujita M, Dresselhaus G and Dresselhaus M S 1996 Edge state in graphene ribbons: nanometer size effect and edge shape dependence *Phys. Rev.* B **54** 17954–61

[10] Son Y-W, Cohen M L and Louie S G 2006 Energy gaps in graphene nanoribbons *Phys. Rev. Lett.* **97** 216803

[11] González J W, Santos H, Pacheco M, Chico L and Brey L 2010 Electronic transport through bilayer graphene flakes *Phys. Rev.* B **81** 195406

[12] Pelc M, Morell E S, Brey L and Chico L 2015 Electronic conductance of twisted bilayer nanoribbon flakes *J. Phys. Chem.* C **119** 10076–84
[13] Hanfland M, Beister H and Syassen K 1989 Graphite under pressure: equation of state and first-order Raman modes *Phys. Rev.* B **37** 12598
[14] Lee J-K, Lee S-C, Ahn J-P, Kim S-C, Wilson J I B and John P 2008 The growth of AA graphite on (111) diamond *J. Chem. Phys.* **129** 234709
[15] Charlier J-C, Gonze X and Michenaud J-P 1994 Graphite interplanar bonding: electronic delocalization and van der Waals interaction *Europhys. Lett.* **28** 403
[16] Palser A H R 1999 Interlayer interactions in graphite and carbon nanotubes *Phys. Chem. Chem. Phys.* **1** 4459–64
[17] Xu Y, Li X and Dong J 2010 Infrared and raman spectra of AA-stacking bilayer graphene *Nanotechnology* **21** 065711
[18] Santos H, Ayuela A, Chico L and Artacho E 2012 van der Waals interaction in magnetic bilayer graphene nanoribbons *Phys. Rev.* B **85** 245430
[19] Sahu B, Min H, MacDonald A H and Banerjee S K 2008 Energy gaps, magnetism, and electric-field effects in bilayer graphene nanoribbons *Phys. Rev.* B **78** 045404
[20] Lopes dos Santos J M B, Peres N M R and Castro Neto A H 2007 Graphene bilayer with a twist: electronic structure *Phys. Rev. Lett.* **99** 256802
[21] Campanera J M, Savini G, Suarez-Martinez I and Heggie M I 2007 Density functional calculations on the intricacies of Moiré patterns on graphite *Phys. Rev.* B **75** 235449
[22] Shallcross S, Sharma S and Pankratov O A 2008 Quantum interference at the twist boundary in graphene *Phys. Rev. Lett.* **101** 056803
[23] Li G, Luican A, Lopes dos Santos J M B, Castro Neto A H, Reina A, Kong J and Andrei E Y 2010 Observation of van Hove singularities in twisted graphene layers *Nat. Phys.* **6** 109
[24] Bistritzer R and MacDonald A H 2010 Moire bands in twisted double-layer graphene *Proc. Nat. Acad. Sci.* **108** 12233
[25] Landgraf W, Shallcross S, Türschmann K, Weckbecker D and Pankratov O 2013 Electronic structure of twisted graphene flakes *Phys. Rev.* B **87** 075433
[26] Sprinkle M *et al* 2009 First direct observation of a nearly ideal graphene band structure *Phys. Rev. Lett.* **103** 226803
[27] Yao W *et al* 2018 Quasicrystalline 30° twisted bilayer graphene as an incommensurate superlattice with strong interlayer coupling *Proc. Natl Acad. Sci.* **115** 6928–33
[28] Hicks J *et al* 2011 Symmetry breaking in commensurate graphene rotational stacking: comparison of theory and experiment *Phys. Rev.* B **83** 205403
[29] Luican A, Li G, Reina A, Kong J, Nair R R, Novoselov K S, Geim A K and Andrei E Y 2011 Single-layer behavior and its breakdown in twisted graphene layers *Phys. Rev. Lett.* **106** 126802
[30] Magaud L, Trambly de Laissardière G and Mayou D 2010 Localization of Dirac electrons in rotated graphene bilayers *Nano Lett.* **10** 804–8
[31] Suárez Morell E, Correa J D, Vargas P, Pacheco M and Barticevic Z 2010 Flat bands in slightly twisted bilayer graphene: tight-binding calculations *Phys. Rev.* B **82** 121407
[32] Suárez Morell E, Vargas P, Chico L and Brey L 2011 Charge redistribution and interlayer coupling in twisted bilayer graphene under electric fields *Phys. Rev.* B **84** 195421
[33] Kim K, DaSilva A, Huang S, Fallahazad B, Larentis S, Taniguchi T, Watanabe K, LeRoy B J, MacDonald A H and Tutuc E 2017 Tunable Moiré bands and strong correlations in small-twist-angle bilayer graphene *Proc. Natl Acad. Sci.* **114** 3364–9

[34] Cao Y *et al* 2018 Correlated insulator behaviour at half-filling in magic-angle graphene superlattices *Nature* **556** 80–4

[35] Cao Y, Fatemi V, Fang S, Watanabe K, Taniguchi T, Kaxiras E and Jarillo-Herrero P 2018 Unconventional superconductivity in magic-angle graphene superlattices *Nature* **556** 43–50

[36] Latil S, Meunier V and Henrard L 2007 Massless fermions in multilayer graphitic systems with misoriented layers: *ab initio* calculations and experimental fingerprints *Phys. Rev.* B **76** 201402

[37] Chico L, Benedict L X, Louie S G and Cohen M L 1996 Quantum conductance of carbon nanotubes with defects *Phys. Rev.* B **54** 2600

[38] Buongiorno Nardelli M 1999 Electronic transport in extended systems: application to carbon nanotubes *Phys. Rev.* B **60** 7828

[39] Katsnelson M 2012 *Graphene: Carbon in Two Dimensions* (Cambridge: Cambridge University Press)

[40] Nilsson J, Castro Neto A H, Guinea F and Peres N M R 2007 Transmission through a biased graphene bilayer barrier *Phys. Rev.* B **76** 165416

[41] Santos H, Chico L and Brey L 2009 Carbon nanoelectronics: unzipping tubes into graphene ribbons *Phys. Rev. Lett.* **103** 086801

[42] Kosynkin D V, Higginbotham A L, Sinitskii A, Lomeda J R, Dimiev A, Price B K and Tour J M 2009 Longitudinal unzipping of carbon nanotubes to form graphene nanoribbons *Nature* **458** 872–6

[43] Jiao L, Zhang L, Wang X, Diankov G and Dai H 2009 Narrow graphene nanoribbons from carbon nanotubes *Nature* **458** 877–80

[44] Cano-Márquez A G, Rodríguez-Macías F J, Campos-Delgado J, Espinosa-González C G, Tristán-López F, Ramírez-González D, Cullen D A, Smith D J, Terrones M and Vega-Cantú Y I 2009 Ex-MWNTs: graphene sheets and ribbons produced by lithium intercalation and exfoliation of carbon nanotubes *Nano. Lett.* **9** 1527–33

[45] Friedman J S, Girdhar A, Gelfand R M, Memik G, Mohseni H, Taflove A, Wessels B W, Leburton J-P and Sahakian A V 2017 Cascaded spintronic logic with low-dimensional carbon *Nat. Commun.* **8** 15635

[46] Klymenko Y O and Shevtsov O 2009 Quantum transport in armchair graphene ribbons: analytical tight-binding solutions for propagation through step-like and barrier-like potentials *Eur. Phys. J.* B **69** 383–8

[47] Huang B, Son Y-W, Kim G, Duan W and Ihm J 2009 Electronic and magnetic properties of partially open carbon nanotubes *J. Am. Chem. Soc.* **131** 17919–25

[48] Bin W and Jian W 2010 First-principles investigation of transport properties through longitudinal unzipped carbon nanotubes *Phys. Rev.* B **81** 045425

[49] Grace I M, Bailey S W and Lambert C J 2004 Electron transport in carbon nanotube shuttles and telescopes *Phys. Rev.* B **70** 153405

[50] Caneva S, Gehring P, García-Suárez V M, García-Fuente A, Stefani D, Olavarria-Contreras I J, Ferrer J, Dekker C and van der Zant H S J 2018 Mechanically controlled quantum interference in graphene break junctions *Nat. Nanotechnol.* **13** 1126–31

[51] González J W, Santos H, Prada E, Brey L and Chico L 2011 Gate-controlled conductance through bilayer graphene ribbons *Phys. Rev.* B **83** 205402

[52] González J W, Pacheco M, Orellana P, Brey L and Chico L 2012 Electronic transport of folded graphene nanoribbons *Sol. State Commun.* **152** 1400–3

[53] Suárez Morell E, Vergara R, Pacheco M, Brey L and Chico L 2014 Electronic properties of twisted bilayer nanoribbons *Phys. Rev.* B **89** 205405
[54] Akhmerov A R and Beenakker C W J 2008 Boundary conditions for Dirac fermions on a terminated honeycomb lattice *Phys. Rev.* B **77** 085423
[55] Jaskólski W, Ayuela A, Pelc M, Santos H and Chico L 2011 Edge states and flat bands in graphene nanoribbons with arbitrary geometries *Phys. Rev.* B **83** 235424
[56] Orellana P A, Rosales L, Chico L and Pacheco M 2013 Spin-polarized electrons in bilayer graphene ribbons *J. Appl. Phys.* **113** 213710
[57] Manchon A, Koo H C, Nitta J, Frolov S M and Duine R A 2015 New perspectives for Rashba spin-orbit coupling *Nat. Mater.* **14** 871–82
[58] Kane C L and Mele E J 2005 Quantum spin Hall effect in graphene *Phys. Rev. Lett.* **95** 226801
[59] Chico L, López-Sancho M P and Muñoz M C 2004 Spin splitting induced by spin-orbit interaction in chiral nanotubes *Phys. Rev. Lett.* **93** 176402
[60] Marchenko D, Varykhalov A, Scholz M R, Bihlmayer G, Rashba E I, Rybkin A, Shikin A M and Rader O 2012 Giant Rashba splitting in graphene due to hybridization with gold *Nat. Commun.* **3** 1232
[61] Santos H, Muñoz M C, López-Sancho M P and Chico L 2013 Interplay between symmetry and spin-orbit coupling on graphene nanoribbons *Phys. Rev.* B **87** 235402
[62] Chico L, Latgé A and Brey L 2015 Symmetries of quantum transport with Rashba spin–orbit: graphene spintronics *Phys. Chem. Chem. Phys.* **17** 16469–75
[63] Qiao Z, Yang S, Feng W, Tse W-K, Ding J, Yao Y, Wang J and Niu Q 2010 Quantum anomalous Hall effect in graphene from Rashba and exchange effects *Phys. Rev.* B **82** 161414
[64] Santos H, Latgé A, Alvarellos J E and Chico L 2016 All-electrical production of spin-polarized currents in carbon nanotubes: Rashba spin-orbit interaction *Phys. Rev.* B **93** 165424
[65] Santos H, Chico L, Alvarellos J E and Latgé A 2017 Defect-enhanced Rashba spin-polarized currents in carbon nanotubes *Phys. Rev.* B **96** 165401

Chapter 6

Electronic transport in graphene nanoribbons

Christoph Tegenkamp, Johannes Aprojanz and Jens Baringhaus

The missing gap in 2D graphene makes it difficult to embed graphene directly in conventional device architectures. Therefore, nanostructures of graphene, and in particular graphene nanoribbons (GNRs), have come into the focus of research over the last few years. Around 1%–2% of the publications dealing with graphene are dedicated to GNRs. Around 1/3 of these papers deal with transport. The propagation of charge carriers directly probes the feasibility of new device concepts and opens the way for the discovery of new states in quantum matter. Besides the possibility to tune the band gap by electron confinement, the edges give rise to robust edge states with fascinating electronic and magnetic properties.

In order to fabricate such well-defined GNRs, various approaches are reported. Besides on-surface synthesis, advanced lithography as well as epitaxial growth experiments have been conducted by many groups. Section 6.3 covers the current state of these major techniques and highlights advantages as well as disadvantages. We will start this chapter by summarizing the fascinating physics expected from ideal GNRs in section 6.1 followed by a brief description of diffusive and ballistic transport in 1D systems, e.g. GNRs (section 6.2). We are aware of further results in this field which we unfortunately could not completely consider here. The aim of this chapter is to provide a representative cross-section, including own work. We apologize for unintentionally missed contributions from our colleagues.

6.1 The role of the edges

The edges not only define the electronic confinement inside the GNR, but they also play an important role in terms of edge states which give rise to new transport phenomena. The lateral confinement of charge carriers in these ribbons of width W leads to an alteration of the band structure which comes along with the opening of a band gap as a consequence of subband formation. According to tight binding calculations, the transverse subband quantization is described by $E_n \approx \pm\hbar v_F\sqrt{k_x^2 + (n\pi/W^2)}$ [1]. The energy gap therefore amounts to $\Delta E \approx 1.9$ eV/W(nm), i.e. for a 40 nm wide ribbon the

doi:10.1088/978-0-7503-1701-6ch6

© IOP Publishing Ltd 2020

(optical) gap is around 50 meV [2]. The energy gap deduced from transport follows $E_g = \Delta E/2$ for an intrinsic system. In terms of GNRs for electronic devices, the band gap is the key parameter. In order to provide an acceptable switch-off for a MOSFET, a band gap of at least 350 meV is mandatory [3]. Neglecting for a moment details of the edge configurations, the width should be below 10 nm in order to reach these requirements which poses a challenge for conventional lithography. This is particularly true if the edge geometry is important. For more details on the performance of GNR MOSFETs, side-gate transistors and three terminal junctions the reader is referred to the following review article [4]. In this section, the focus will be more on the implications of atomistic details in the edge geometries.

The honeycomb lattice favors edge orientations with cuts parallel to the C–C bond forming so-called armchair GNRs (ac-GNR). In contrast, a zigzag GNR (zz-GNR) is rotated by 30° with respect to the armchair geometry. Cutting in any other direction results in a ribbon with a mixed armchair and zigzag edge, called chiral GNR (see figure 6.1). A huge difference between an ac-GNR and zz-GNR is the origin of the edge atoms. While the armchair edge has alternating edge atoms from both sublattices, the edge atoms in a zigzag edge all belong to the same sublattice. In case of an ac-GNR, the edge symmetry demands the vanishing of the wave function for both sublattices (A, B) at both edges, which gives rise to three classification regimes for ac-GNRs. Only those with $N_a = 3p$, $3p + 1$, where p is a positive integer, are semiconducting. N_a = 2 W/0.243 nm is the number of C-atoms across the ribbons. The variation of the gap size is well understood and confirmed by STS measurements, e.g. done on epitaxially grown ac-GNRs (see section 6.3.3) [1, 5, 6].

Similarly, the energy spectrum for zz-GNR can be calculated, giving rise to similar bulk band gaps and subband variations depending on the width [7]. Up to

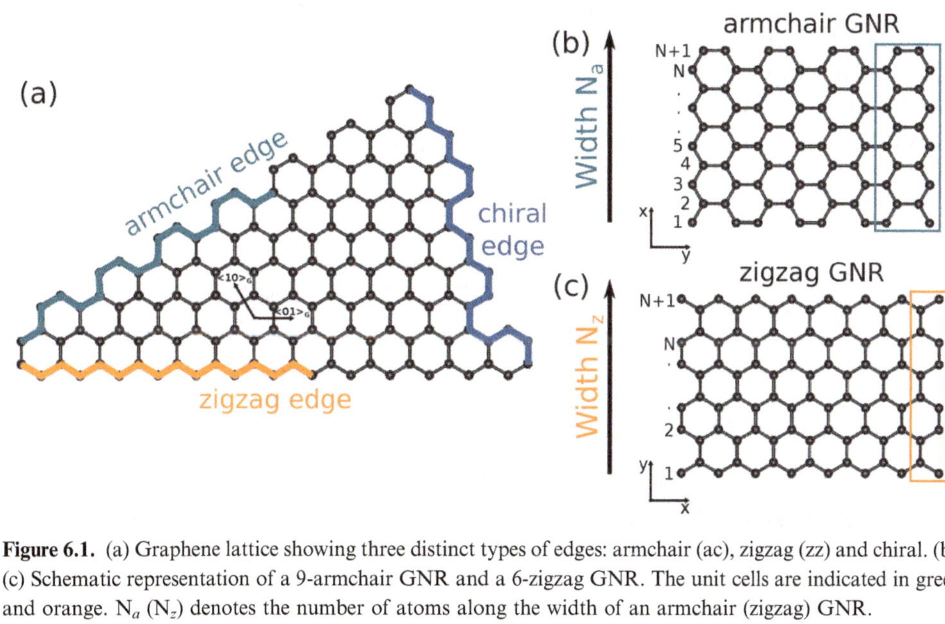

Figure 6.1. (a) Graphene lattice showing three distinct types of edges: armchair (ac), zigzag (zz) and chiral. (b), (c) Schematic representation of a 9-armchair GNR and a 6-zigzag GNR. The unit cells are indicated in green and orange. N_a (N_z) denotes the number of atoms along the width of an armchair (zigzag) GNR.

this, the energy spectra for GNRs are the very same as compared to carbon nanotubes (CNTs) [8] (note the unzipping of zz-CNT results in an ac-GNR and vice versa). However, the zz-GNRs provide some special characteristic compared to all other carbon sp^2-allotropes: the lift of the AB-lattice structure demands the existence of a metallic edge state [7, 9, 10]. Both the valence and conduction band edge state merge at E_D. They stay flat roughly in between $2\pi/3a \leqslant k \leqslant \pi/a$ and this flattening even increases with increasing ribbon width, as sketched in figure 6.2. Thereby, the location of the electrons for various k-values of this range changes from localization at the edges towards delocalization for $k = \pi/a$ [11]. Due to the flat bands, the DOS of the zz-GNR exhibits a sharp peak at E_F, in contrast to the vanishing DOS in 2D graphene. These flat bands at zero energy should give rise to so-called perfectly conducting channels (PCCs) [12].

However, such singularities at the Fermi level are prone to instabilities [1, 13, 14]. By theory, various new ground states of zz-GNRs were proposed. First-principles calculations showed that charge ordering and Peierls-like distortions in zz-GNRs of reasonable width ($N_z \approx 50(10\ \text{nm})$) are energetically unfavorable. Instead, a remarkably long-ranged magnetic interaction across the ribbons gives rise to spin-polarized edge states [15]. For the antiferromagnetic edge order, a band gap opens, thus the zz-GNR becomes a so-called Slater insulator [1, 16, 17]. However, for ribbons wider than 7 nm the antiferromagnetic coupling collapses and the band gap vanishes [18]. In contrast, in the case of ferromagnetic edge coupling, the spin-resolved edge states interpenetrate and no band gap is present, hence the zz-GNR is always metallic (Stoner metal) independent of the width [19]. Accordingly, the edge states mimic spin-polarized Dirac states of a topological insulator and the charge carriers propagating along the ribbon are protected against backscattering as long as time reversal symmetry is preserved (see below). For both kinds of ribbons, the magnetic effects on the bulk bands are negligibly small [15]. With increasing ribbon width, the bulk bands approach and finally recover graphene's band structure. The edge magnetism, however, is not affected. These considerations only hold true as long as freestanding ribbons with H-termination are considered. Any bonding to the substrate or asymmetric termination of the edges opens a path for further peculiar

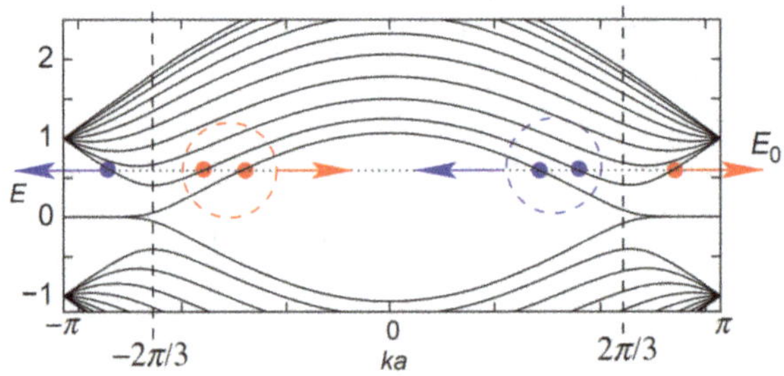

Figure 6.2. Band structure of a 10-zz-GNR. The red (blue) circles denote the right (left) moving transport channels for an electron doped ribbon ($E_F = E_0 > E_D = 0$). In the left (right) valley, the degeneracy between the left and right moving channels is missing due to one excess right (left) propagating channel. Reproduced from [11]. © IOP Publishing Ltd and Deutsche Physikalische Gesellschaft. All rights reserved.

properties. For instance, a transverse electric field comes along with half-metallicity for antiferromagnetically coupled zz-GNRs [13]. Since the spins at opposite edges are oriented in opposite directions, the electric field shifts the occupied and unoccupied states of one spin closer together and those of the other type further apart. In case of closing the gap, the ribbon becomes half-metallic and the transport is spin-polarized.

Often, in realistic systems, where the GNR interacts with its support, the planarity is lifted by strain fields despite the strong in-plane σ-bonds. Strong curvature gives rise to strong, but time-reversal symmetric pseudomagnetic fields and the observation of zero-field Landau levels in graphene [20, 21]. For GNRs, it was shown that the inclusion of spin–orbit coupling within the edge states mimics a Zeeman-type term, thus pseudomagnetic effects can in principle also affect the spin degree of freedom, e.g. giving rise to a quantum spin Hall state with helical spin channels [22].

This brief recapitulation on GNRs shows that the control of the edge symmetry is decisive in order to control the charge and spin degree of freedom and to understand electronic correlation effects [23]. However, the regimes rely on perfect edges and symmetry. Any edge misalignment introduces kinks giving rise to chiral ribbons [24–26]. Nonetheless, the edge states are robust against disorder. Perturbations that break the electron–hole symmetry, e.g. edge impurities or the hopping integral, shift the edge state away from zero energy but do not alter their total density [27].

In the limit of perfect edges at zero temperature, the electrical conductance in a GNR follows the Landauer formula $G(E) = 2e^2/h \times g(E)$, where $g(E)$ denotes the number of subbands crossing E_F (see equation (6.1); for further details, see section 6.2) [28]. Neglecting electron–electron interactions and spin polarization effects of the edge states, $g(E) = n$, $n + 1$ and $2n + 1$ for semiconducting ac-GNRs, metallic ac-GNRs and zz-GNRs, respectively. Generally, disorder will lead to localization of the propagating electrons making it very challenging to observe ballistic transport in GNRs. Thereby, in each of the parabolic subbands inter- as well as intravalley scattering are possible depending on momentum transfer provided by defects. However, as is obvious from figure 6.2, for the zeroth subband, intravalley back-scattering is not possible. Therefore, in the left (right) valley there is always one excess right (left) moving channel showing the ballistic conductance of $2e^2/h$. Only upon the introduction of short-range impurities, this PCC vanishes as now intervalley scattering is enabled. For further reading, we refer to [11]. Depending on the distribution of the defects, instead of ballistic conductance a diffusive transport regime or even Coulomb blockade effects may occur.

6.2 Transport regimes in nanoribbons: from diffusive to Coulomb blockade to ballistic

The transport properties of any nanodevice depend fundamentally on the regime in which electrons are conducted through the device. For GNRs, three main regimes shall be considered here: diffusive and ballistic transport as well as Coulomb blockade. An overview of these regimes is provided in figure 6.3. The discrimination between diffusive and ballistic transport is straightforward considering two characteristic length scales, the electron mean free path λ_e and the phase coherence length λ_Φ. While the mean free path describes the distance which an electron can travel without experiencing a scattering event, the phase coherence length can be understood as the distance over which quantum coherence of the electron ensemble is preserved. Electrons can travel ballistically if both their mean free path and phase coherence length largely exceed the lateral dimensions of the nanodevice, i.e. the GNR in our case. The electrons undergo neither inelastic nor elastic scattering events and are only reflected from the boundaries of the structure and hence, its shape controls the electron trajectories. A special situation occurs if due to disorder transport in nanodevices no longer happens continuously but rather between temporarily electrically disconnected islands. This situation is typically observed in GNRs obtained by lithographic etching (see section 6.3.2). Due to edge roughness, which causes a variation of the width along the GNR, the number of conducting channels is reduced and electrons are temporarily confined in individual quantum dot-like islands. This leads to the manifestation of a transport gap since the electron motion between these islands suffers from Coulomb blockade effects [29].

Diffusive transport in which electrons are elastically and inelastically scattered while traveling through the device is well described by using the classical theory of Drude. Here, the effect of scattering events is summarized in the mobility of the system. Any external bias on the contacts to the device causes an electric field which accelerates the electrons linearly, depending on the mobility, and induces a corresponding drift current as shown in figure 6.4(a). Hence, the conductivity of the system is simply the product of electron density, charge and mobility: $\sigma = \rho^{-1} = en\mu$. In the classical Drude theory, all conduction electrons are participating in the drift current. In DC transport measurements on diffusive systems, the resistance comprises an intrinsic part due to scattering of electrons in the sample and parasitic contact resistances due to the probes. The latter is almost negligible if instead of a two point-probe (2pp) a four point-probe (4pp) configuration is used with high impedance voltage probes [30].

This is fundamentally different at low temperatures ($k_BT \ll E_F - E_0$), where only electrons close to the Fermi energy can contribute to the current. By increasing both the mean free path and phase coherence length well above the spatial dimensions of the GNR, electrons enter into the ballistic transport regime. Consequently, the band structure of the electric conductor transits from continuous to inhibiting only a few

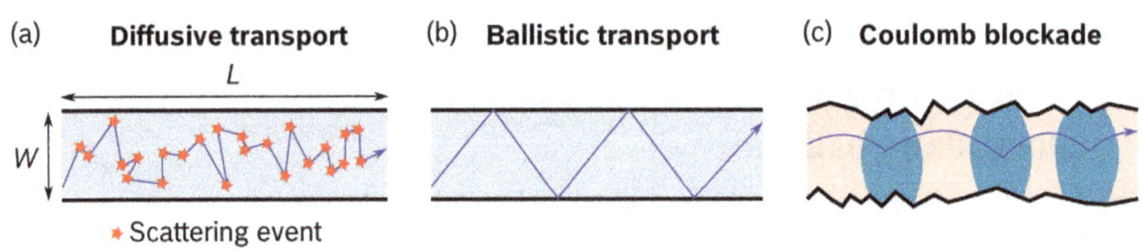

Figure 6.3. Schematic illustration of electron trajectories in the (a) diffusive, (b) ballistic and (c) Coulomb blockade transport regime.

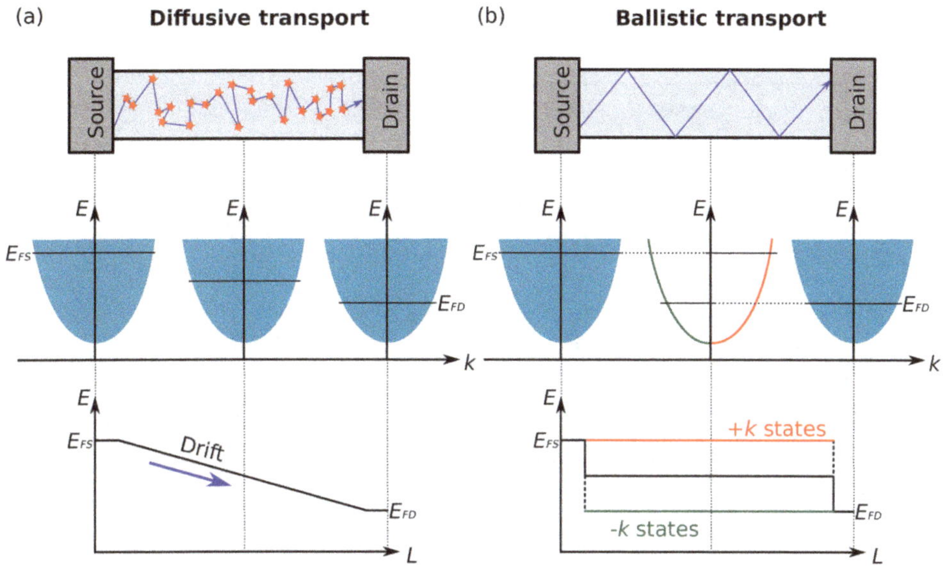

Figure 6.4. Schematic illustration of a conductor with two ohmic contacts and their corresponding dispersion relation and band diagrams. The case of a diffusive conductor is shown in (a), while (b) depicts a ballistic conductor. The location of quasi-Fermi levels at the source (E_{FS}) and drain contact (E_{FD}) is given in both dispersion relations and band diagrams.

modes; figure 6.4(b). The description of ballistic transport outlined in the following is based on the theory of Datta [28]. The application of a bias voltage to a ballistic conductor leads to a splitting of the quasi-Fermi levels of both contacts. Assuming perfect contacts with no reflection occurring at the interface, electrons enter from the contact of a higher energy level, travel ballistically, hence with no loss of energy, through the conductor and enter the energetically lower contact. At this point, a voltage drop has to occur in order to enter the contact. This voltage drop can be directly associated with a contact resistance. The same argument applies for electrons traveling in the other direction (into the energetically higher contact). Since the voltage drops at the interfaces to the contacts are the only ones occurring, a ballistic conductor is said to feature only contact but no intrinsic resistance. A property which can also be understood in direct analogy to the aforementioned Drude model, where the resistivity decreases with increasing mobility. The ballistic conductor is, in this analogy, a system with infinite mobility due to the absence of scattering events. Its conductance can be calculated by summing over left and right moving electrons using Fermi–Dirac statistics. The total conductance G is quantized with the number of modes, i.e. subbands M. Imperfections along the channels are considered by a channel-specific transmission probability T_n, so that the famous Landauer expression reads:

$$G = \frac{2e^2}{h} \sum_{n=1}^{M} T_n. \tag{6.1}$$

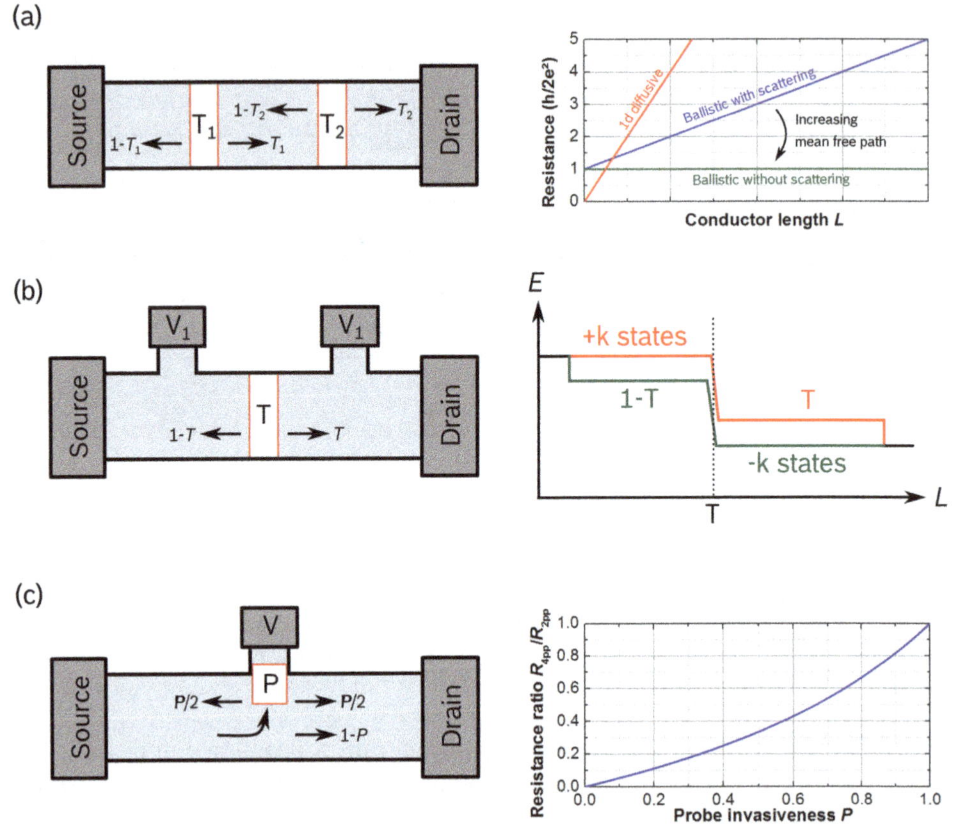

Figure 6.5. Schematic illustration of different scenarios of ballistic transport in conductors inheriting scattering centers. (a) The case of two scattering centers T_1 and T_2 is shown as well as the impact of an infinite number of scattering centers with $T \approx 1$ on the length dependence of the resistance. The influence of a single scattering center T in a four-point probe setup is depicted in (b) together with the band diagram illustrating the change of quasi-Fermi levels. The situation with an invasive voltage probe (P) is shown in (c) alongside the resistance ratio between a four- and a two-point probe measurement caused by invasive probes.

The Landauer expression can be used to describe the influence of individual scatterers within the ballistic device. As sketched in figure 6.5(a) for the case of two scattering centers, the net transmission through these scatterers is obtained by summation over their respective transmission and reflection coefficients. Assuming a large number of scattering centers with low reflection ($T \approx 1$), the resistance of the ballistic conductor can be expressed as a function of its length L [28]:

$$R(L) = \frac{h}{2e^2M}\left(1 + \frac{L}{\lambda_e}\right). \tag{6.2}$$

It is found that the resistance is linearly increasing with the length of the ballistic conductor reaching a value of $\frac{h}{2e^2}$ when extrapolated to $L = 0$. Hence, a simple two-point probe transport measurement with variable probe spacing can be effectively used to characterize the presence of scattering centers, the corresponding mean free path as well as the ballistic nature of the GNR. A similar resistance measurement carried out on a one-dimensional diffusive conductor yields a resistance linearly depending on the length (or probe spacing): $R = \rho L/(Wt)$ [31], and extrapolates to zero (see figure 6.5). From the slope, the 1D resistivity ρ/Wt can be read off directly.

Extending the electrical characterization of a nanodevice from a simple two-point probe measurement to a four-point probe setup leads to the surprising result of vanishing resistance in ballistic systems [32]. This can be understood by considering the electrochemical potentials during a four-point probe measurement. The setup is illustrated in figure 6.5(b) for the case of a ballistic conductor including one scattering center with transmission probability T. The voltage probes which are free of current measure a difference of $1 - T$ in the normalized electrochemical potential which consequently leads to a resistance R_{4pp} depending on the transmission probability T.

$$R_{4pp} = \frac{h}{2e^2 M} \frac{1 - T}{T}. \tag{6.3}$$

Hence, in the absence of scattering ($T = 1$) the resistance measured in a four-point probe setup is zero. This can also be understood in analogy to a four-point probe measurement performed on a diffusive conductor. Of course, this applies only to perfect voltage probes, i.e. they do not interact with the device under test and do not interfere with the electron flow. On the other hand, typical voltage probes used on nanodevices can have a finite probability with which an electron flowing inside the conductor can enter the voltage probe. After entering the voltage probe, they are reinjected into the conductor and their phase information is lost [33, 34]. This probability is the invasiveness of the probe P and its impact on transport within the conductor is depicted in figure 6.5(c). A fully invasive voltage probe can be treated like a scattering center with a transmission probability of $T = 1/2 (P = 1)$. Consequently, the resistance measured when two invasive voltage probes are present between the current carrying contacts can be calculated by considering the transmission and reflection coefficients between the two probes [35, 36]. The ratio between the resistance of a ballistic conductor measured in a four-point probe and a two-point probe setup then reads:

$$\frac{R_{4pp}}{R_{2pp}} = \frac{P}{2 - P}. \tag{6.4}$$

In the case of $P = 1$, both the four-point and the two-point probe measurement give the same result which can be understood from the simple picture of fully invasive voltage probes intercepting the conduction and hence splitting up the conductor into three equal pieces [32]. Moreover, using ballistic systems, the probes can be characterized by determining their invasiveness (see figure 6.5(c)). By varying the contact material and contacting pressure both invasive and non-invasive probes were realized on GNRs [37].

6.3 Transport studies on GNRs

This section will summarize and discuss recent results and advances obtained in the field of electronic transport in GNRs. In literature, three strategies are mainly followed. Ultra-small ribbons below 10 nm with atomically precise edges are obtained by surface-assisted chemical reactions of benzene-based precursor molecules [38, 39]. Recently, these concepts were successfully also applied to insulating surfaces, e.g. TiO_2, so that the parasitic conductivity of metallic substrates, e.g. Au (111), is not limiting envisaged transport measurements [40]. Section 6.3.1 will give a brief overview of the recent progress. Besides such bottom-up approaches, top-down strategies are pursued. Thereby, subtractive etching techniques were gradually optimized over the last few years resulting in better-defined edges [41]. Generally, using exfoliated graphene allows us to sandwich it between other layers, e.g. h-BN or other 2D materials, thus providing a flexible way for further functionalization. Some pioneering work in this field is highlighted in section 6.3.2. Thirdly, self-assembled growth of ribbons by so-called sublimation epitaxy on pre-patterned SiC-surfaces can be used to grow ac-GNRs as well as zz-GNRs [42]. For the latter, ballistic transport was reported for the first time at room-temperature [43]. Section 6.3.3 will summarize the latest work in this field. There are many more exciting examples, which cannot be all addressed here. Without making complete claims, we exemplarily mention here the unzipping of CNTs [44], anisotropic etching of graphene by nanoclusters [45] or wrinkle structures within 2D graphene grown on Ni(111) [46].

6.3.1 GNRs formed by on-surface synthesis

The realization of well-shaped and ultra-narrow GNRs succeeds by means of surface assisted-coupling of appropriate precursor molecules on metallic surfaces. This allows an atomically precise bottom-up fabrication of GNRs. The prototypical 10,10'-dibromo-9,9'-bianthryl (DBBA) molecules undergo on Au(111) Ullmann reactions forming straight ac-GNRs [38, 47]. An ac-GNR with $N_a = 9$ based on 3',6'-dibromo-1,1':2',1'-terphenyl precursors is shown in figures 6.6(a) and (b) [48]. Moreover, the catalytic activity is tunable by the substrate, e.g. for DBBA on Cu (111) the Ullmann coupling is inactive and chiral GNRs are formed [49]. Moreover, chevron-type GNRs with different degree of n-doping were successfully synthesized by employing tetraphenyltriphenylene-based monomers with different degrees of n-doping [50]. The on-surface synthesis has been further optimized, in a way that even hetero-junctions, e.g. with MnPc to ac-GNRs, can be realized [51]. A comprehensive overview of the chemistry of nanographene can be found in [52].

As mentioned above, most of the precursor molecules result in the growth of straight, chiral or meandering GNR structures with armchair-type edges. The synthesis of zz-GNR, which are expected to reveal robust metallic edge states, is by far more difficult. From a chemical point of view, using so-called U-shaped benzene-based monomers with different functional groups, i.e. halogens as well as biphenyl-ligands, provided a breakthrough in the field of on-surface synthesis [17]. Such monomers undergo on Au(111) polymerization by aryl-aryl coupling at 475 K. At the cyclodehydrogenation temperature of around 625 K, this snake-like polymer

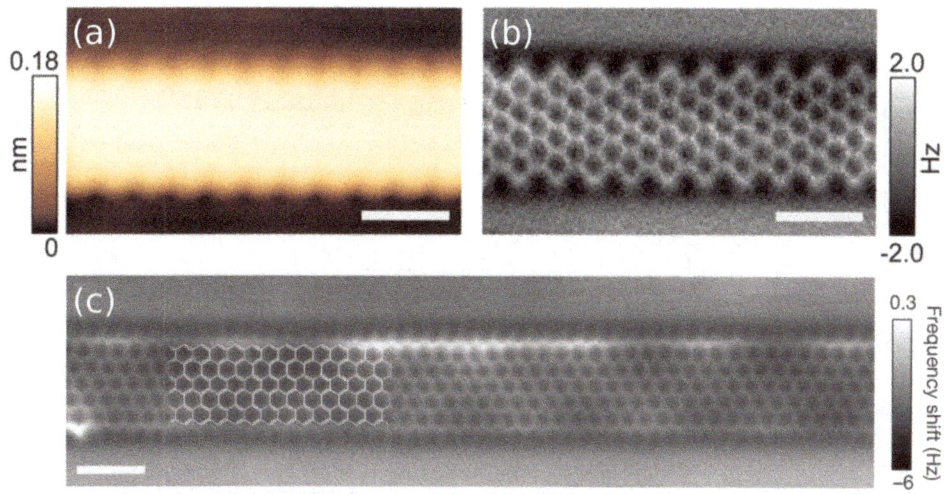

Figure 6.6. (a) High-resolution STM topography image of a single ac-GNR ($N_a = 9$, $U = 0.1$ V, $I = 0.5$ nA). (b) High-resolution nc-AFM frequency shift image of ac-GNR ($N_a = 9$) using a CO-functionalized tip with an oscillation amplitude $A_{osc} = 70$ pm [48]. (c) Constant height nc-AFM frequency shift image taken with a CO-functionalized tip. The intra-ribbon resolution shows the formation of a zz-GNR ($N_z = 6$) with atomically precise CH edges. A CH_2 defect is seen in the lower left corner ($A_{osc} = 70$ pm, $U = 5$ mV) [17]. Scale bars 1nm. (a), (b) Reprinted with permission from [48]. Copyright (2017) American Chemical Society. (c) Reprinted by permission from Nature [17], © 2016 Springer Nature Publishing AG.

finally transforms into a zz-GNR with $N_z = 6$ as shown in figure 6.6(c). Indeed, by collecting conductance maps with STS, the edge states could be clearly identified, like on other systems [24]. In agreement with theory, the edge-edge interaction lifts the degeneracy at E_F and opens a gap of around 1.5 eV for 6-zz-GNR (≈11 Å) [1, 16]. However, absolute values depend strongly on residual interactions with the substrate [23, 26]. Weakly coupled GNRs can be realized, e.g. by manipulating the ribbons onto small NaCl islands [39]. Most recently, direct growth of GNRs on rutile titania surfaces succeeded on the basis of carbon–fluorine bonds to enable aryl–aryl coupling schemes [40]. So far, transport in a 2-terminal configuration was performed for GNRs synthesized from DBBA on Au(111) [38]. Their width is as small as 7.4 Å, thus revealing strong quantum confinement and a semiconducting behavior ($N_a = 7 = 3p + 1$, $p = 2$).

Scanning tunneling microscopy (STM) is a powerful technique as it allows us to control the atomic shape of the ribbons, as well as the spatial mapping of their electronic structure. Moreover, by controlled pulling experiments with a STM-tip, GNRs can be lifted in an extremely controlled way [53]. By variation of the tip-height, length-dependent 2-terminal transport measurements can be performed under ultra-clean conditions at low temperatures (10 K) [54]. The $I(z)$-curves shown in figure 6.7(a) clearly reveal an exponential decay which is indicative for electron tunneling along the GNRs. Depending on the overall length of the GNR, a certain tip-height z refers to a characteristic effective length L (see the inset of figure 6.7(a)). The conductance of the ac-GNR for this assembly follows in this case

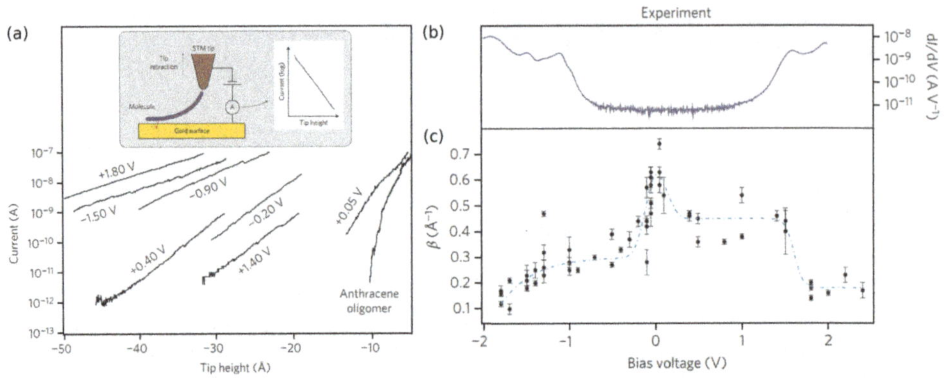

Figure 6.7. (a) Current as a function of tip height ($I(z)$) for different experiments (zero tip height refers to tip surface contact) at fixed bias voltages. (Inset) Schematic of the STM pulling experiment with an ac-GNR (arrow indicates tunneling current). A characteristic current signal during the pulling sequence is shown in the right panel. (b) dI/dV spectrum for a nanoribbon in a pulling geometry (fixed length). (c) The attenuation factors β, which are obtained from the $I(z)$-curves shown in (a) plotted as a function of the bias voltage. A dashed line is drawn to guide the eye. The experimental β values, determined from many nanoribbons, need to be reduced (by $\approx$ 10%–15%) to obtain the real values because the STM tip height is slightly smaller than the effective molecular length in the junction. Reprinted with permission from Nature Nanotechnology [54], © 2012 Springer Nature Publishing AG.

$G(L) = I/V = G_0 e^{-\beta L}$, where G_0 is the quantum point conductance and β the inverse decay length [55]. The $I(z)$-signals follow perfectly this theoretical expectation and show no significant noise, which underlines the stability of the molecule as well as of the contacts during the transport measurement. The decay length depends on the bias voltage, i.e. on the position of the electronic states in the GNR. The dI/dV-signal for fixed tip-height, shown in panel (b), clearly exhibits the HOMO- and LUMO-states at −1.1 V and +1.6 V, respectively. Using graphene as a reference, the gap size of 2.7 eV can be seen as the result of strong confinement within this 7 Å wide ribbon [6]. This molecular structure is reflected in the attenuation factor β: less extended states for tunneling across the ribbon lead to a higher inverse decay length. In agreement with theory, the GNRs are gapped and reveal no edge states. Indications of these were seen at the termini of the ac-GNRs. However, for the long ribbons, there is no overlap of the two so-called Tamm states located at each of the ends of the ac-GNR. A more recent study has investigated this effect in more detail [39]. The asymmetry in β is due to the asymmetric distribution of the electronic states (figure 6.7(c)). Most severely, the inverse decay length shows a peak close to the Fermi level E_F. This effect originates from the bending of the lifted ac-GNR during STM manipulation. Thus, a ballistic transport across the ac-GNR (bulk states) is expected only for planar GNRs in energetic resonance with the molecular states [54].

6.3.2 GNRs made by lithography

GNRs may provide a sufficiently large band gap and are therefore considered to be building blocks for graphene-based electronics [3]. Towards applications, the ability to realize multi-terminal setups, e.g. with a gate electrode, is essential. These requirements can be realized proceeding from graphene and subsequent use of lithography. Thereby, CVD grown graphene, epitaxial graphene on (insulating) SiC as well as exfoliated graphene flakes are well-established templates [57–59]. In the following, we will concentrate mainly on GNRs made on the basis of exfoliated graphene. Such GNRs can be flexibly combined with other 2D materials for protecting ribbon edges and offering further routes for functionalization.

A multi-terminal contact setup based on Cr/Au (3/50 nm) on exfoliated graphene is shown in figures 6.8(a) and (b). The GNR itself can be defined as small as 10 nm by an e-beam lithography process, e.g. based on hydrogen silsesquioxane (HSQ). Exemplarily, the conductance G is shown in (c) for a 24 nm wide ribbon. By varying the (global) gate, an ambipolar transport behavior is demonstrated and the conductance curve reveals a transport gap ΔV_g, which is clearly present at low temperatures. If C_g is the capacitive coupling of the GNR to the back gate, the energy gap of the GNR can be deduced via $E_g = \hbar v_F \sqrt{2\pi C_g \Delta V_g / |e|}$ [56, 61]. For

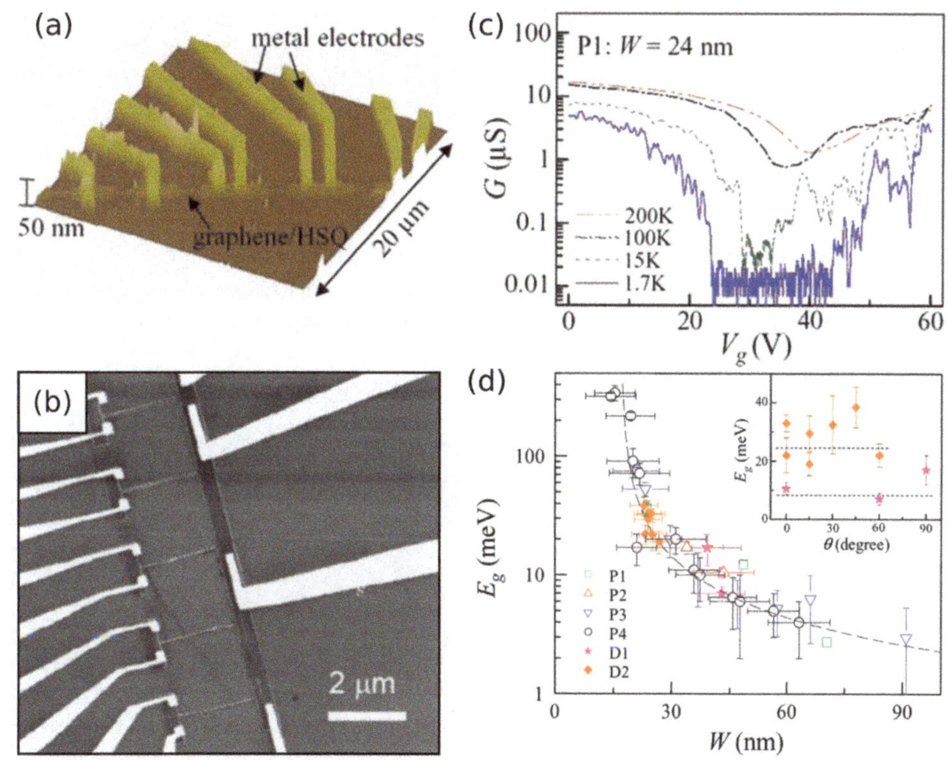

Figure 6.8. (a) Atomic force microscope image of a GNR (device P3) covered by a protective HSQ etch mask. (b) SEM image of device set P1 with parallel GNRs of varying width. (c) Conductance of a GNR ($W = 24$ nm) in device set P1 as a function of gate voltage measured at different temperatures. (d) Gap energy E_g versus the GNR width W for six device structures. Reprinted with permission from [56], Copyright (2007) by the American Physical Society.

various GNRs of different widths, the results summarized in (d) show clearly the expected $E_g \propto \Delta E \propto 1/W$ behavior. Outside of the gapped region, the conductance along the ribbon is proportional to the back gate ($G \propto V_g$, see figure 6.9(a)), which is distinctive for the diffusive transport regime [60, 62].

However, the gap feature is strongly temperature dependent and points towards defect states which are easily introduced during the GNR fabrication process [63]. Moreover, the minimum in $G(V_g)$-plots indicates the charge neutrality point, hence the ribbons seem to be intrinsically doped, most likely from resist residuals or due to charge transfer from the dense array of contacts. Sometimes, gap sizes significantly differ from the expected values ($E_g \gg 1eV/W$(nm)) [59]. Including disorder, different energy scales found in these systems can be explained in terms of Coulomb blockade effects [59, 64]. This means that the gap measured is rather a so-called mobility or transport gap and, moreover, edge effects as seen in the previous section are entirely quenched.

The quality of graphene was strongly improved by its encapsulation with hexagonal BN (h-BN), revealing charge carrier mobilities up to $\mu = 10^5 \mathrm{cm}^2\, \mathrm{V}^{-1}\mathrm{s}^{-1}$ at room temperature. Further passivation by another planar 2D material enables also the realization of back and top gate coupling [65, 66]. At least for graphene nanoconstrictions, shown as an inset in figure 6.9(b), this technique improved the transport characteristics. According to the Landauer formula, a square root dependence of the conductance as a function of V_g is indicative for ballistic quantum transport ($G \propto \sqrt{V_g}$) [67]. Moreover, superimposed on this, the conductance reveals kink structures of nearly G_0 which are referred to ballistic transport contributions from bulk channels [60] (see figures 6.9(b) and (c)). Despite the great progress towards transport and also the improvement of lithography [41], edge control on the atomic scale is still not realized for GNRs. First indications of atomically precise edge termination were recently observed in magnetotransport measurements carried

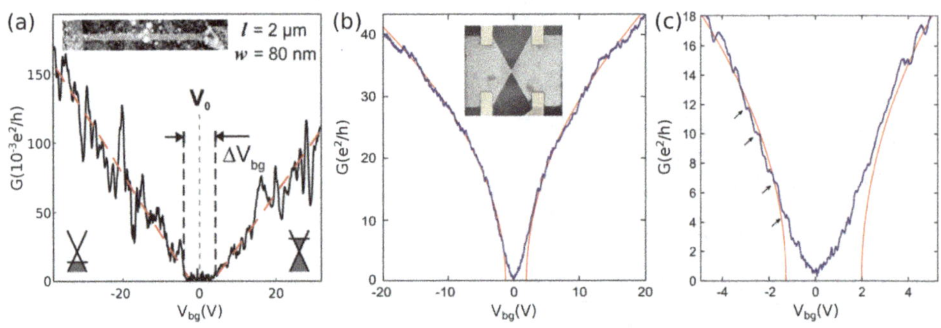

Figure 6.9. (a) Two-terminal back gate characteristics of a 80 nm wide GNR shown in the inset. $G \propto V_g$ is indicative for diffusive transport. (b) Conductance trace of a 280 nm wide encapsulated graphene nanoconstriction (inset) as a function of the back gate voltage (blue trace) taken at $T = 4$ K. (c) Close-up of the trace in (b) around the CNP. Conductance kinks in steps of nearly G_0 as denoted by the black arrows are observed. $G \propto \sqrt{V_g}$ is obtained for a ballistic system. Reprinted from [60] John Wiley & Sons. Copyright © 2017 WILEY-VCH Verlag GmbH & Co. KGaA, Weinheim.

out on 70 nm wide ribbons, where a zeroth-order Landauer conductance peak was found only for zz-GNR [68].

Recently, helium-ion beam etched GNRs were investigated. Compared to the conventional lithographic steps, the nanostructuring of (encapsulated) graphene can be performed *in situ* without the need for resists. The good focusing performance of a He-ion microscope (HIM) allows the fabrication of arbitrarily shaped, sub-10 nm wide GNRs [69]. Conductance measurements at low temperatures reveal an energy gap, which indeed scales inversely with the ribbon width. However, despite the good spatial resolution of HIM, the edges are still rough giving rise to diffusive transport and also the observation of Coulomb blockade effects [70].

6.3.3 GNRs epitaxially grown on SiC

A third class of ribbons are so-called epitaxial graphene ribbons (epi-graphene ribbons). Compared to the previous methods, here the growth of graphene or its intercalation is the final step so that it is not corrupted by subsequent processing. In the following, we will highlight three examples. First, so-called natural ribbons at SiC step edges are presented [71, 72]. The mono- and bilayer structures are electronically decoupled from the SiC(0001) substrate. The high mobilities in these ribbons having comparably rough edges rely on interlayer hopping of the charge carriers, thus effectively bypassing defects present in each of the layers. Secondly, pn-type junctions were fabricated by intercalation of Ge on pre-structured buffer layer structures on SiC(0001) [73, 74], which is also an interesting approach for future transistor applications. A final example covers so-called sidewall ribbon structures, where the SiC(0001) surface is lithographically structured prior to growth. The subsequent annealing favors the growth of graphene on the edges of the SiC-mesas [42, 75]. The GNR-edges are well defined on these mesas and reveal outstanding transport properties.

Natural edge ribbons: diffusive transport

The annealing of SiC at high temperatures comes along with the effective desorption and sp^2-hybridization of Si and C, respectively. On vicinal SiC(0001) substrates with steps running along the $[11\bar{2}0]$ direction, this can be utilized to grow arrays of parallel and non-interconnected stripes of graphene in a controlled manner. Thereby, in a first step monolayer (ML) GNRs are formed with an electron concentration of $n \approx 1 \times 10^{13}\mathrm{cm}^{-2}$ (middle panel of figure 6.10(a)). Moreover, the recently proposed *ex situ* oxidation step nicely forms p-type doped bilayer GNRs, which are electronically disconnected from the SiC template [71]. The typical width of the ribbons varies between 30–100 nm. The result of this selective growth process is sketched in figure 6.10(a) and structural details can be found in [72].

Contacting individual natural GNRs using lithographically made contacts is quite challenging. Instead, by using an *in situ* multiprobe technique (4-tip STM/SEM), the tips can be precisely navigated to the desired positions and allow length dependent 4-terminal measurements. Such approaches substitute the fabrication of multiple fixed contacts as exemplarily shown in figure 6.8(b). The accidental

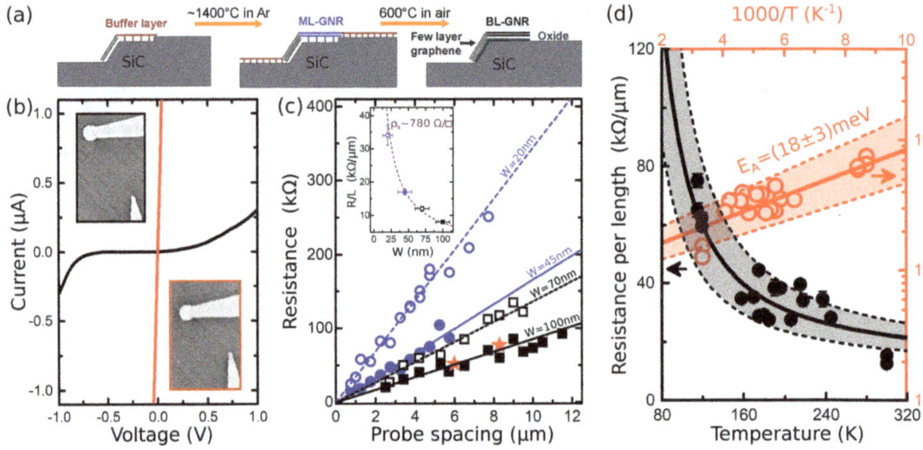

Figure 6.10. (a) Schematic for the growth of monolayer (ML) and bilayer (BL) GNRs at step edges of SiC. (b) Characteristic two-point-probe I(V) curves taken on the ribbon (linear, red) and with one of the two tips accidentally contacted to the SiC-substrate (Schottky-diode like, black). (c) Resistance as a function of probe spacing for various natural GNRs of different width and thickness t (ML-GNR, BL-GNR) measured both with the 2 point probe (2pp) and collinear 4 point probe (4pp, red stars) geometries. The inset shows that all GNRs reveal the sheet resistance of 780Ω/□. All measurements were performed at 300 K. (c) The resistance per length R/L of various BL-GNRs having comparable width, as a function of temperature. The same data points are also plotted on the logarithmic scale against the inverse temperature (i.e., Arrhenius plot) and clearly show a temperature activated transport behavior in BL-GNRs. (We thank A A Zakharov, Lund, for providing the ML-GNR sample.)

placement of one of the tips besides the ribbon is directly reflected in a Schottky-like behavior of the measured current (see figure 6.8(b) for reference). Besides a different transport characteristic, the absolute current value is also about three orders of magnitude smaller than the value measured on the nanoribbons at similar bias voltages. This demonstrates that the natural GNRs are indeed electronically decoupled from the SiC-substrate.

Figure 6.8(b) shows the resistance as a function of probe spacing. It is large compared to the width W, for ML-GNRs and BL-GNRs of different widths. The resistance increases linearly with the two-point probe spacing which is an expected signature for 1D diffusive transport [31]. The linear regression intercepts the origin, i.e. the contact resistances of the tips are negligible. The two data points (marked by stars) were instead measured in a four-point probe geometry and underlines that the cleaning protocols of the tips and their gentle approach results in extremely low contact resistances.

The resistance per length R/L corresponding to the two BL-GNRs is around 12 kΩ/μm and 8 k$\Omega/\mu m$, respectively. From SEM images (see insets in figure 6.8(b)), we deduced an average width of $W \approx 70$ nm and $W \approx 100$ nm, respectively, which yields comparable sheet resistances of $\rho_s = (R/L) \times W \approx 800\Omega/\square$ for these ribbons, comparable to other diffusive graphene systems [72]. Taking the hole concentration in these ribbons (induced by the intercalation of oxygen) into account [71, 76], this resistivity corresponds to an effective hole mobility of around $\mu = 700$ cm^2 V^{-1}s^{-1}.

As the mobility and carrier concentration are known, these values result in a mean free path length of $\lambda_e = \hbar\mu\sqrt{n\pi}/e \approx 3$ nm [77]. Moreover, the ML-GNRs, of similar charge carrier concentration (but electrons instead of holes), reveal a very similar sheet resistance (see inset of figure 6.10(c)). This estimation shows that the atomic roughness of the edges is irrelevant for the transport along the natural ribbons.

More insight is gained from temperature dependent measurements. The results of the resistance per length as a function of temperature are summarized in figure 6.8(d) for various BL-GNRs of similar widths. The scattering of the data basically reflects the slightly different GNR width. Nonetheless, there is a clear trend that the resistance per length decreases by a factor of four during heat up from 100 K to room temperature. This behavior is surprising if compared to single-layer graphene, which typically shows a metallic behavior induced by the weak electron–phonon interaction [78]. For example, for epitaxial graphene on SiC (i.e. for the ML-GNRs), the mobilities are increased only by around 40% due to remote phonon scattering [79]. The effect seen for the BL-GNRs is much stronger and shows the opposite trend. From the $\log(R)$ versus $1/T$ plot (see figure 6.8(c)), an activation energy of 18 ± 3 meV is deduced. AB-stacked BLG is expected to reveal a band gap, which could be of this size including electric field effects originating from the interface. However, in our case, the BL-GNRs are heavily p-type doped [76], thus interband transitions can be excluded. The band gap in these ribbons is at maximum around 10 meV. The π-orbitals of the graphene are extended symmetrically with respect to the plane of the hexagonally arranged C-atoms. While in MLG the electronic transport is restricted to this channel, the propagation in BLG occurs along the π-bands located around each of the two layers. Therefore, interlayer hopping becomes feasible [80]. The importance of phonon-assisted interlayer tunneling was pointed out recently [81] and this mechanism also determines the conductivity in graphite in the direction across the basal planes [82]. Moreover, the interlayer hopping of the charge carriers is helpful in order to circumvent in-plane defects. Indeed, line defects, which cross the entire ribbon, were observed in recent phase-contrast AFM experiments and are apparently caused by the process of intercalation. The activation energy of 18 ± 3 meV corresponds to the double excitation of a layer breathing and shear phonon mode (also called ZO' + C mode) in bilayer graphene [83]. Apparently, the combination of the shearing and layer breathing modes increases the wave function overlap of the π-states of the same sublattices in adjacent layers in a constructive manner in AB-stacked graphene. This scenario increases the hopping probability of the charge carriers along the wires.

Epitaxial ribbons with ballistic pn-junctions: Klein tunneling

Besides the electronic-gap related ideas for GNR electronic devices [3], there are other effects in graphene which could enable novel transistor concepts, e.g. Klein tunneling [84, 85]. A building block of this concept is the bipolar junction which advances future electron optics, e.g. lenses and beam splitters or Fabry–Pérot interferometers [86, 87]. Appropriate bipolar transistor structures (npn- and pnp-junctions) were proposed to be realized by locally manipulating the chemical potential such that the height of the potential barrier V_0 exceeds the energy of the

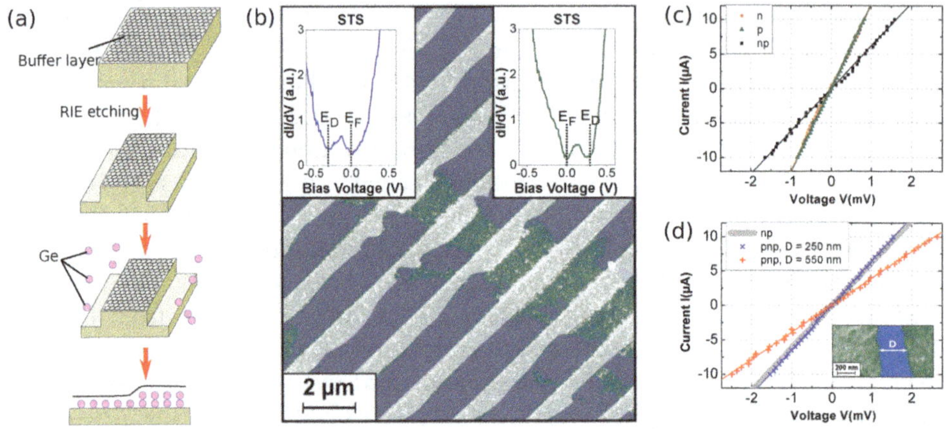

Figure 6.11. (a) Schematic of the different processes (formation of the buffer layer, etching, and intercalation) in order to reveal GNRs of defined width W with spatially different chemical potentials. (b) Color-coded SEM image. The green (blue) colors denote p-type (n-type) graphene areas. The local chemical potentials are deduced from local spectroscopy (STS) curves shown as insets. (c) IV-measurement (collinear 4pp) across a single pn-junction at $T = 30$ K. As reference, also the IV-curves measured on n- and p-type samples are shown. (d) pnp-junctions for different barrier lengths D. The insets show corresponding SEM images of the junctions. For the short barrier lengths, the IV-curves fit perfectly into the gray-shaded areas, which represent the values measured across a single pn-junction ($T = 30$ K) [74].

electrons obeying to a linear dispersion. In a purely one-dimensional scenario, these relativistic electrons will fully transmit across the heterojunction irrespective of the barrier lengths, which is confined in between perfect pn-junctions [84]. However, any experimental signature of Klein tunneling depends on the underlying transport regimes and, in particular, on the barrier characteristics, i.e. the pn-junction and the barrier length D.

Transport experiments across bipolar graphene junctions can be realized by electric field effects due to appropriately designed top gate structures. However, electrostatic stray fields present at the edges of the gate contacts cause the pn-junctions to be typically of 1 μm in size giving rise to so-called smooth junctions [88]. In the case of diffusive graphene, the transmitted electrons may undergo scattering within the pn-junction. Alternatively, ultra-narrow pn-junctions down to 5 nm and well-defined potential barriers can be realized by means of intercalation [74]. In particular, n- and p-type doped graphene can be realized by monolayer and bilayer of intercalated Ge, respectively [73, 89, 90]. A big advantage of using epitaxially grown buffer layer structures on SiC(0001) templates is the long-range ordering of the buffer layer which is maintained after intercalation. Thus, transport across the junctions is not limited by disorder [91, 92]. Figure 6.11(a) schematically shows the fabrication process of n- and p-type doped ribbons based on pre-etched buffer layer structures [93]. The width of the ribbons here is around 1 μm, i.e. band gap formation due to confinement does not play a role.

The combined STM-SEM microscopy enables us to correlate the different intensity levels with p- or n-type doped graphene areas and, therefore, allows us to navigate and approach the nanoprobes reliably to desired positions for surface sensitive transport measurements (see figure 6.11(b) [89]). The STS spectra (insets in (b)) show the double minimum structure characteristic for intact graphene in which E_F is not coinciding with E_D. Exactly two levels of chemical potentials evolve which are directly correlated with mono- or bilayer thick Ge-films underneath [89]. The electron concentration in the n-type area ($n = 8 \times 10^{12} \text{cm}^{-2}$) is slightly higher than the hole concentration ($p = 6 \times 10^{12} \text{cm}^{-2}$) in the p-type areas which is in perfect agreement with previous results obtained by angle resolved photoemission experiments [73].

In the following, the transmission across these junctions are probed by four-point probe transport measurements. The transport in each of the subsystems is diffusive irrespective of the type of doping. From temperature dependent measurements of the sheet resistance, the mobilities of both n- and p-type areas at low temperatures (30 K) are around $\mu \approx 2000 \text{ cm}^2 \text{ V}^{-1}\text{s}^{-1}$. Considering the carrier concentrations in each of the subsystems, this refers to an elastic mean free path length of $\lambda_e \approx 100$ nm, which is 20 times larger than the length of the pn-junction [74]. The use of an *in situ* nanoprobe system allows us to measure the resistances of the junctions separately. The IV-curve recorded at low temperature across the pn-junction is shown in figure 6.11(c). For comparison, the IV-curves recorded on the n- and p-type areas are also plotted and are virtually identical. The total resistance across a single pn-junction at 30 K deduced from the IV-curves is $R_{pn} = U/I = (162 \pm 5)\ \Omega$ while the resistance in each of the subsystems (either n- or p-type) measures $R_{n(p)} = (84(82) \pm 4)\ \Omega$.

All IV-curves are taken in the same collinear fashion and the distance of the inner probes from the barrier is approximately 750 nm, i.e. large compared to the mean free path length. The resistance of a single pn-junction itself can be calculated simply via $R_{pn-junc} = R_{pn} - R_{n(p)} = (78(80) \pm 3)\ \Omega$. As a consequence of the diffusively propagating electrons towards the barrier, only the charge carriers with a momentum inside the acceptance window of the Klein-transmission curve can be transmitted. All other charge carriers are reflected and contribute to the resistance. Based on the structural and electronic parameters ($k_F = 0.037\ \text{Å}^{-1}$), the junction with a nominal length of $t = 5$ nm can be classified as $k_F t < 1$ ($k_F t \approx 0.7$) so that the resistance is in reasonable agreement with theoretical expectations for a sharp junction, $R_{pn}^{sharp} = 3h\pi/8e^2 k_F W = 82\ \Omega$ [94, 95].

Knowing the tunneling characteristics of a single pn-junction, transport across npn- and pnp-junctions with variable barrier lengths D in between the two junctions can be measured. In analogy to optics, the first pn-junction takes up the part of a polarizer while the second is the analyzer. The Ge intercalated samples, as shown in figure 6.11(b), provide various barrier lengths (for both polarities) between $D = 200$–500 nm. The corresponding IV-curves across npn- and pnp-junctions are shown in figure 6.11(d). For barrier lengths around 200 nm, where $D \approx 2\lambda_e$, the total resistance is identical to the resistance measured across a single pn-junction, i.e. the resistance of the second junction is fully transparent. This does not depend on the polarity and is the same for npn- and pnp-junctions. Obviously, those electrons filtered by the first barrier transmit fully through the second junction [74]. For longer barriers, the electrons randomize and experience reflectance of the second barrier. This example shows the formation of ballistic Klein tunneling barriers in a globally diffusive environment simply by intercalation of Ge in long-range ordered buffer layer structures grown on SiC templates.

Epitaxial sidewall ribbons: towards ballistic conductors
Except for the surface-assisted growth of ultra-narrow ribbons presented in section 6.3.1, the transport characteristics are not influenced by details of the edge geometry. To date, subtractive methods, as shown in section 6.3.2, are not suitable to fabricate GNRs with defined edges. The results obtained over recent years show that epitaxial graphene grown on mesa-structured SiC(0001) turns out to be a suitable template to grow GNRs of defined widths with defined edge orientations. Based on the epitaxial relation between graphene and SiC(0001), mesa-structures running along the $[1\bar{1}00]$- and $[11\bar{2}0]$-directions trigger the formation of zz-GNRs and ac-GNRs, respectively [42].

The selective growth of graphene on SiC-sidewalls at temperatures of 1800 °C competes with the formation of new SiC facets as well as the different etching rates of the different types of steps and moreover depends on the SiC-polytypes [96, 97]. So far, the growth conditions were optimized for 6H-SiC, which allow wafer-scale growth of edge specific GNR structures [75]. Figure 6.12 summarizes SEM and STM results showing the selective growth of ac-GNRs and zz-GNRs. Thereby, only for mesas running along the $[1\bar{1}00]$-directions, stable $\{11\bar{2}n\}$-facets with an average inclination angle of $25 \pm 2°$ are formed on which extended ribbons growth can be found (figure 6.12(b)). In contrast, the mesa walls for ac-GNR growth consist of a number of microsteps (10–20) of 2–3 nm ($N_A \approx 20$) in width and 1–1.5 nm in height (see figure 6.12(c)). The ac-GNRs appear to be strongly corrugated following the topography of the substrate; its whole width is larger (up to 100 nm) with regards to the zz-GNR, and the sidewall slope is slightly smaller. It should be noted that similar types of nanorippled ac-GNRs can be also fabricated via sublimation epitaxy on vicinal SiC(0001) surfaces [98, 99].

The difference in the average widths of the two kinds of ribbons is reflected in the STS spectra. The zz-GNR show still the semimetallic dI/dV signal of a neutral and extended graphene ribbon with no signs of confinement. The ac-GNR shows a gap of around 300–400 meV, which refers to a ribbons width of ≈ 5 nm. Systematic STS measurements done on ac-GNRs grown on facets of SiC(0001) and SiC($000\bar{1}$) surfaces show an almost perfect agreement with DFT calculations [1, 5, 6]. This is also reflected in the four-point probe transport experiments. The zz-GNR shows a resistance close to 1 h/e^2, while the resistance of the armchair GNR is almost 10 times larger at the same probe spacing. These results prove that carefully grown narrow GNRs behave as conventional semiconductors and can be considered as potential candidates for electronic devices [3]. Albeit STS locally showed a semiconducting gap, with the lateral transport averaging up to 10 ac-GNRs. Moreover, transport paths along defects (steps) are probed, thus the local transport measurement shows an increased linear resistance rather than a clear signature of

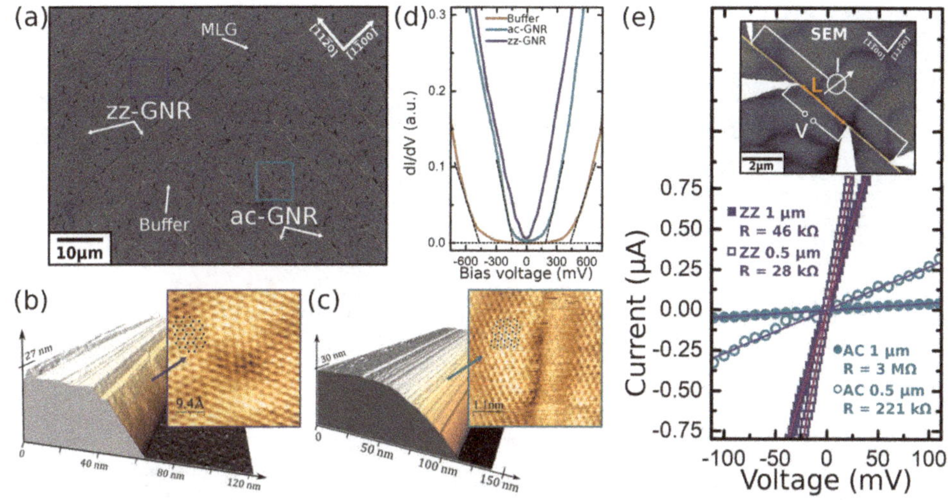

Figure 6.12. (a) SEM image showing the area with mesas hosting ac-GNRs and zz-GNRs. Besides small patches of 1 Ml graphene, there is almost no variation of the SEM contrasts levels across the entire surface, demonstrating a remarkable homogeneity of the buffer layer. (b), (c) STM 3D representations of a zz-GNR and ac-GNR. The insets show magnifications. (d) STS spectra taken on the buffer layer (BL), the ac-GNR and zz-GNR showing semiconducting (setpoint +1 V, 500 pA) and semimetallic (+1.3 V, 70 pA) IV-characteristics, respectively. (e) *In-situ* four-point probe transport measurements of zz-GNR and ac-GNR. Exemplary IV curves taken for two distinct probe spacings L on both kinds of GNR. The zz-GNR shows a resistance $\approx h/e^2$, while the resistance of the ac-GNR is almost 10 times larger at the same probe spacing. The collinear tip arrangement is depicted in the inset. Reproduced from [75].

semiconducting behavior. The semiconducting properties of ac-GNRs are obtained for atomically precise edges following quantization rules. If, for instance, chirality is included, these GNRs can also be metallic.

Most interestingly, the sidewall zz-GNRs show a finite quantum resistance of h/e^2, which is a result of the edge geometry and termination and resembles a ballistic conductor [37, 43, 93, 97, 100–102]. In contrast to the ac-GNRs, the STS shown in figure 6.13(b) show clear edge states, in qualitative agreement with the GNR band structure discussed in the context of figure 6.2 [93]. The spectra taken towards the center show a gap. However, this gap Δ_{elastic} is not a result of electronic confinement but rather arise from suppression of electronic tunneling to graphene states near E_F. The simultaneous enhancement of electronic tunneling at higher energies is due to activation of an optical phonon ($\approx$ 60 meV) [103, 104], thus this gap-like feature is a characteristic fingerprint of a suspended GNR.

Figure 6.13(a) shows the collinear contact assembly of four STM tips brought into an ohmic contact to the zz-GNR. The outer and inner probes are used as current and voltage probes, respectively. The resistance is plotted in (c) as a function of different probe spacings L on different zz-GNRs on the sample. All measurements follow $R(L) = h/e^2(1 + L/\lambda_e)$ (see equation (6.2)). The intersection at h/e^2 is a hallmark for a ballistic system. Moreover, the 'missing' factor of 2 suggests that the spin-degeneracy is lifted and that the propagating electrons are spin-polarized. The

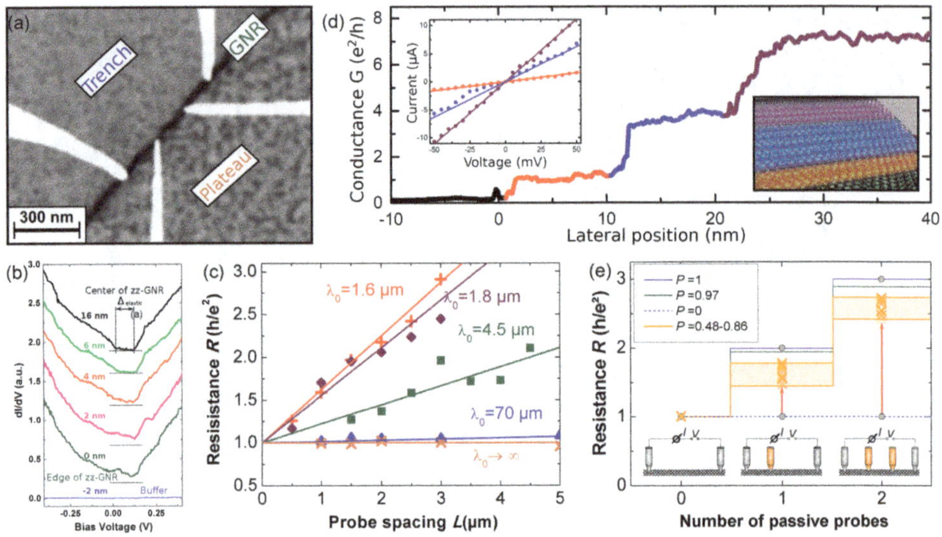

Figure 6.13. (a) SEM image of four collinearly arranged STM tips placed on a zz-GNR. (b) Sequence of STS spectra taken at different positions across the zz-GNR. (c) 4pp resistance measurements as a function of the probe spacing. All curves extrapolate to h/e^2. (d) 2pp conductance measurements for probe spacings of 70 nm. The transverse movement of one of the tips opens sequentially the edge channel (red) and bulk channels (blue, purple). (e) Passive probe experiment: the orange-colored probes in the sketch act as (almost) perfect scattering centers and double/triple the quantum resistance. All measurements are performed at 300 K. Reproduced from [43, 93, 100].

slope reflects the mean free path lengths λ_e. The systematic investigations by our group have revealed that λ_e correlates with the average terrace lengths of the SiC (0001) surfaces, thus the zz-GNR interacts with the SiC surface [105]. In case of an extremely high SiC step density, the zz-GNR shows a diffusive behavior ($R \propto L$) [101], while the most perfect ribbons show an exceptional channel with mean free paths of up to 17 μm at 300 K [43, 97].

In a recent experiment, spatially resolved transport measurements were performed. Using a 2 terminal geometry with ultra-small spacings of 70 nm, one extremely sharp tip was moved transversely across the ribbon from the lower towards the upper edge. The other blunt tip, covering the entire 40 nm wide ribbon, was kept fixed in position. As obvious from figure 6.13(d), first the e^2/h edge channel is detected followed by the appearance of bulk channels revealing the 4-fold spin- and pseudo-spin degeneracy of graphene [100]. Compared to the bulk states with a mean free path of $\lambda_{\text{bulk}} \approx 3\pi W/4 = 95$ nm [106], the edge state is much more robust ($\lambda_e \gg \lambda_{\text{bulk}}$). Moreover, in agreement with electron interference experiments using nanoconstrictions realized by *in situ* STM lithography at the lower part of the zz-GNR [102] and conductive AFM measurements [100], the edge channel is located at the bottom part of the sidewall ribbon. Thirdly, the spatial separation of the bulk channels, seen in figure 6.13(d), is a result of an asymmetric edge bonding, giving rise to a transverse electric field, as confirmed by tight-binding calculations [100]. This shows that the bonding of at least the zigzag edges to the substrate is extremely important and introduces new functionality. Moreover, the bonds to the substrate protect the electronic properties and the epitaxial sidewall ribbons are much less prone to chemical modifications. The ballistic channel can be measured even under ambient conditions [100].

In context of the idealized band structure in figure 6.2 for a zz-GNR, the spin-degeneracy of the edge channel is lifted here, giving rise to e^2/h. Assuming a shift of the states like in case of ferromagnetic edge coupling, the situation becomes similar to that of a topological insulator, where backscattering is prohibited as long as time reversal symmetry is preserved. We have to admit that details about the electronic structure of the edge states are unknown to date and that further investigations are mandatory.

Finally, the role of the contacts shall be elucidated in more detail. Both 4pp- and 2pp-measurements shown in figures 6.13(c) and (d) reveal the same conductance. This is due to the fact that the nanoprobes can act in a fully invasive way. As outlined in section 6.2, the R_{4pp} in a ballistic system depends on the invasiveness P of the voltage probes. For instance, a scattering center with a certain transmission T in a ballistic system causes a potential drop of the propagating electrons and the resistance reads $R_{4pp} = R_{2pp}(1 - T)/T$ (see equation (6.4)). As is obvious, in the absence of a scatter (T = 1), the resistance vanishes. In the second extreme case, scattering centers with equal transmittance and reflectance of $T = R = 1 - T = 0.5$ will result in identical $2pp$ and $4pp$ resistance values [32, 37, 107]. Being aware of this, the ballistic nature of the edge channel can be easily proven. Starting with a 2 terminal configuration, the h/e^2-contact is measured. The other two probes act now as so-called passive probes and simply act as scatterers. Accordingly, the total inverse transmission is doubled and tripled if one and two passive probes are brought in between, shown in figure 6.13(e), for various invasiveness parameters P. As expected for ballistic systems, the switching of contact resistances is not dependent on the position of the passive probes [43].

6.4 Summary and conclusion

We have recapped here some of the latest results about electronic transport in GNRs. The control of the GNR width, the edge smoothness, and the edge orientation are decisive to make use of the tunable band gaps for transistor applications or edge channels for the dissipation-less propagation of electrons. The physics behind the role of the edges in GNRs and some relevant transport aspects were outlined in sections 6.1 and 6.2.

Currently, three approaches are the focus of research: the on-surface synthesis, lithography and epitaxial growth as outlined in section 6.3. The most narrowest GNRs of a few Å with structurally perfect and chemically intact zigzag or armchair edge configurations are made by surface chemistry, as explained in section 6.3.1. Thereby, the chemical reactions are triggered by the catalytic activity of the (metallic) surfaces, which at the same time limits this approach for subsequent multi-terminal transport measurements. Unless STM-based 2 terminal experiments are insufficient, transfer techniques to dielectric supports are mandatory.

Alternatively, new on-surface synthesis routes on insulating surfaces, e.g. TiO_2 [40], are promising. The lithographic and etching approach is in terms of pattering and structuring an extremely versatile and flexible approach, also with respect to new 2D materials other than graphene (section 6.3.2). However, this approach still suffers from rough edges and, instead, the observation of mobility gaps and Coulomb blockade effects. Nonetheless, the development in lithography allows a better edge control [41]. Also, helium ion microscopy is employed to a less destructive and well-controlled formation of ribbon structures out of 2D materials [70, 108].

The fabrication of epitaxial ribbons was outlined in section 6.3.3. Compared to the former techniques, here the growth of graphene was done finally after the SiC template was structured. However, the growth process is rather delicate. Due to the high temperatures required for sublimation, the SiC surface itself may undergo refacetting. So far, the sidewall ribbons grown on 6H-SiC revealed ballistic transport. The origin of the robust edge channel is still not yet understood on quantum mechanical grounds. Compared to the on-surface synthesized GNRs with C-H edges, the termination for the sidewall ribbons is very different. One characteristic seems to be the formation of σ-bonds between the GNR and the SiC. The edge hybridization may protect the GNRs, but at the same time curvature effects arise coming along with transverse electric field effects [100].

In conclusion, there is strong progress in the field of transport in GNRs. The concept of asymmetric edge termination may open new ways to design new states of quantum matter in graphene nanostructures. However, in order to understand the transport properties, microscopy on an atomic scale is indispensable.

References

[1] Son Y-W, Cohen M L and Louie S G 2006 Energy gaps in graphene nanoribbons *Phys. Rev. Lett.* **97** 216803

[2] Shen H, Shi Y and Wang X 2015 Synthesis, charge transport and device applications of graphene nanoribbons *Synth. Met.* **210** 109–22

[3] Schwierz F 2010 Graphene transistors *Nat. Nanotechnol* **5** 487

[4] Geng Z, Hhnlein B, Granzner R, Auge M, Lebedev A A, Davydov V Y, Kittler M, Pezoldt J and Schwierz F 2017 Graphene nanoribbons for electronic devices *Ann. Phys.* **529** 1700033

[5] Palacio I *et al* 2015 Atomic structure of epitaxial graphene sidewall nanoribbons: flat graphene, miniribbons, and the confinement gap *Nano Lett.* **15** 182–9

[6] Wang W-X, Zhou M, Li X, Li S-Y, Wu X, Duan W and He L 2016 Energy gaps of atomically precise armchair graphene sidewall nanoribbons *Phys. Rev.* B **93** 241403

[7] Wakabayashi K, Sasaki K-I, Nakanishi T and Enoki T 2010 Electronic states of graphene nanoribbons and analytical solutions *Sci. Technol. Adv. Mater.* **11** 054504

[8] Avouris P, Dresselhaus M S and Dresselhaus G (ed) 2001 *Carbon Nanotubes* (Berlin: Springer)

[9] Nakada K, Fujita M, Dresselhaus G and Dresselhaus M S 1996 Edge state in graphene ribbons: nanometer size effect and edge shape dependence *Phys. Rev.* B **54** 17954–961

[10] Wakabayashi K, Fujita M, Ajiki H and Sigrist M 1999 Electronic and magnetic properties of nanographite ribbons *Phys. Rev.* B **59** 8271–82

[11] Wakabayashi M Y K, Takane Y and Sigrist M 2009 Electronic transport properties of graphene nanoribbons *New J. Phys.* **11** 095016
[12] Wakabayashi K, Takane Y and Sigrist M 2007 Perfectly conducting channel and universality crossover in disordered graphene nanoribbons *Phys. Rev. Lett.* **99** 036601
[13] Son Y-W, Cohen M L and Louie S G 2006 Half-metallic graphene nanoribbons *Nature* **444** 347
[14] Dutta S and Pati S K 2010 Novel properties of graphene nanoribbons: a review *J. Mater. Chem.* **20** 8207–23
[15] Pisani L, Chan J A, Montanari B and Harrison N M 2007 Electronic structure and magnetic properties of graphitic ribbons *Phys. Rev.* B **75** 064418
[16] Yang L, Park C-H, Son Y-W, Cohen M L and Louie S G 2007 Quasiparticle energies and band gaps in graphene nanoribbons *Phys. Rev. Lett.* **99** 186801
[17] Ruffieux P *et al* 2016 On-surface synthesis of graphene nanoribbons with zigzag edge topology *Nature* **531** 489
[18] Magda G Z, Jin X, Hagymási I, Vancsó P, Osváth Z, Nemes-Incze P, Hwang C, Biró L P and Tapasztó L 2014 Room-temperature magnetic order on zigzag edges of narrow graphene nanoribbons *Nature* **514** 608–11
[19] Huang L F, Zhang G R, Zheng X H, Gong P L, Cao T F and Zeng Z 2013 Understanding and tuning the quantum-confinement effect and edge magnetism in zigzag graphene nanoribbon *J. Phys.: Condens. Matter* **25** 055304
[20] Guinea F, Katsnelson M I and Geim A K 2009 Energy gaps and a zero-field quantum Hall effect in graphene by strain engineering *Nat. Phys.* **6** 30
[21] Levy N, Burke S A, Meaker K L, Panlasigui M, Zettl A, Guinea F, Castro Neto A H and Crommie M F 2010 Strain-induced pseudo-magnetic fields greater than 300 Tesla in graphene nanobubbles *Science* **329** 544
[22] He W-Y and He L 2013 Coupled spin and pseudomagnetic field in graphene nanoribbons *Phys. Rev.* B **88** 085411
[23] Li Y Y, Chen M X, Weinert M and Li L 2014 Direct experimental determination of onset of electron–electron interactions in gap opening of zigzag graphene nanoribbons *Nat. Commun.* **5** 4311
[24] Merino P, Santos H, Pinardi A L, Chico L and Martin-Gago J A 2017 Atomically-resolved edge states on surface-nanotemplated graphene explored at room temperature *Nanoscale* **9** 3905–11
[25] Jia X, Campos-Delgado J, Terrones M, Meunier V and Dresselhaus M S 2011 Graphene edges: a review of their fabrication and characterization *Nanoscale* **3** 86–95
[26] Tao C *et al* 2011 Spatially resolving edge states of chiral graphene nanoribbons *Nat. Phys.* **7** 616
[27] Wimmer M, Akhmerov A R and Guinea F 2010 Robustness of edge states in graphene quantum dots *Phys. Rev.* B **82** 045409
[28] Supriyo D 2000 *Electronic Transport in Mesoscopic Systems* (Cambridge: Cambridge University Press)
[29] Sols F, Guinea F and Castro Neto A H 2007 Coulomb blockade in graphene nanoribbons *Phys. Rev. Lett.* **99** 166803
[30] Wenner F 1915 A method of measuring earth conductivity *Bull. Bur. Stand.* **12** 469
[31] Miccoli I, Edler F, Pfnr H and Tegenkamp C 2015 The 100th anniversary of the four-point probe technique: the role of probe geometries in isotropic and anisotropic systems *J. Phys.: Condens. Matter* **27** 223201

[32] de Picciotto R, Stormer H L, Pfeiffer L N, Baldwin K W and West K W 2001 Four-terminal resistance of a ballistic quantum wire *Nature* **411** 51–4

[33] Büttiker M 1986 Role of quantum coherence in series resistors *Phys. Rev.* B **33** 3020–26

[34] Brouwer P W and Beenakker C W J 1995 Effect of a voltage probe on the phase-coherent conductance of a ballistic chaotic cavity *Phys. Rev.* B **51** 7739–43

[35] Büttiker M 1986 Four-terminal phase-coherent conductance *Phys. Rev. Lett.* **57** 1761–64

[36] Ihn T 2010 *Semiconductor Nanostructures* (Oxford: Oxford University Press)

[37] Aprojanz J, Miccoli I, Baringhaus J and Tegenkamp C 2018 1D ballistic transport channel probed by invasive and non-invasive contacts *Appl. Phys. Lett.* **113** 191602

[38] Cai J *et al* 2010 Atomically precise bottom-up fabrication of graphene nanoribbons *Nature* **466** 470–4

[39] Wang S, Talirz L, Pignedoli C A, Feng X, Mllen K, Fasel R and Ruffieux P 2016 Giant edge state splitting at atomically precise graphene zigzag edges *Nat. Commun.* **7** 11507

[40] Kolmer M, Zuzak R, Steiner A K, Zajac L, Engelund M, Godlewski S, Szymonski M and Amsharov K 2019 Fluorine-programmed nanozipping to tailored nanographenes on rutile TiO_2 surfaces *Science* **363** 57

[41] Jessen B S *et al* 2019 Lithographic band structure engineering of graphene *Nat. Nanotechnol.* **14** 340–6

[42] Sprinkle M, Ruan M, Hu Y, Hankinson J, Rubio-Roy M, Zhang B, Wu X, Berger C and de Heer W A 2010 Scalable templated growth of graphene nanoribbons on SiC *Nat. Nanotechnol.* **5** 727

[43] Baringhaus J *et al* 2014 Exceptional ballistic transport in epitaxial graphene nanoribbons *Nature* **506** 349–54

[44] Kosynkin D V, Higginbotham A L, Sinitskii A, Lomeda J R, Dimiev A, Price B K and Tour J M 2009 Longitudinal unzipping of carbon nanotubes to form graphene nanoribbons *Nature* **458** 872–76

[45] Campos L C, Manfrinato V R, Sanchez-Yamagishi J D, Kong J and Jarillo-Herrero P 2009 Anisotropic etching and nanoribbon formation in single-layer graphene *Nano Lett.* **9** 2600–04

[46] Lim H, Jung J, Ruoff R S and Kim Y 2015 Structurally driven one-dimensional electron confinement in sub-5-nm graphene nanowrinkles *Nat. Commun.* **6** 8601

[47] Kimouche A, Ervasti M M, Drost R, Halonen S, Harju A, Joensuu P M, Sainio J and Liljeroth P 2015 Ultra-narrow metallic armchair graphene nanoribbons *Nat. Commun.* **6** 10177

[48] Talirz L *et al* 2017 On-surface synthesis and characterization of 9-atom wide armchair graphene nanoribbons *ACS Nano* **11** 1380–88

[49] Schulz F *et al* 2017 Precursor geometry determines the growth mechanism in graphene nanoribbons *J. Phys. Chem.* C **121** 2896–904

[50] Cai J *et al* 2014 Graphene nanoribbon heterojunctions *Nat. Nanotechnol.* **9** 896

[51] Su X, Xue Z, Li G and Yu P 2018 Edge state engineering of graphene nanoribbons *Nano Lett.* **18** 5744–51

[52] Narita, Wang X-Y, Feng X and Müllen K 2015 New advances in nanographene chemistry *Chem. Soc. Rev.* **44** 6616–43

[53] Temirov R, Lassise A, Anders F B and Tautz F S 2008 Kondo effect by controlled cleavage of a single-molecule contact *Nanotechnology* **19** 065401

[54] Koch M, Ample F, Joachim C and Grill L 2012 Voltage-dependent conductance of a single graphene nanoribbon *Nat. Nanotechnol. Lett.* **7** 713

[55] Lafferentz L, Ample F, Yu H, Hecht S, Joachim C and Grill L 2009 Conductance of a single conjugated polymer as a continuous function of its length *Science* **323** 1193

[56] Han M Y, Özyilmaz B, Zhang Y and Kim P 2007 Energy band-gap engineering of graphene nanoribbons *Phys. Rev. Lett.* **98** 206805

[57] Hwang W S, Tahy K, Li X, Xing H G, Seabaugh A C, Sung C Y and Jena D 2012 Transport properties of graphene nanoribbon transistors on chemical-vapor-deposition grown wafer-scale graphene *Appl. Phys. Lett.* **100** 203107

[58] Hwang *et al* 2015 Graphene nanoribbon field-effect transistors on wafer-scale epitaxial graphene on SiC substrates *APL Mater.* **3** 011101

[59] Molitor F, Stampfer C, Güttinger J, Jaobsen A, Ihn T and Ensslin K 2010 Energy and transport gaps in etched graphene nanoribbons *Semicond. Sci. Technol.* **25** 1

[60] Somanchi S, Terrs B, Peiro J, Staggenborg M, Watanabe K, Taniguchi T, Beschoten B and Stampfer C 2017 From diffusive to ballistic transport in etched graphene constrictions and nanoribbons *Ann. Phys.* **529** 1700082

[61] Adam S, Cho S, Fuhrer M S and Das Sarma S 2008 Density inhomogeneity driven percolation metal-insulator transition and dimensional crossover in graphene nanoribbons *Phys. Rev. Lett.* **101** 046404

[62] Puddy R K, Chua C J and Buitelaar M R 2013 Transport spectroscopy of a graphene quantum dot fabricated by atomic force microscope nanolithography *Appl. Phys. Lett.* **103** 183117

[63] Melinda Y,, Han J, Brant C and Kim P 2010 Electron transport in disordered graphene nanoribbons *Phys. Rev. Lett.* **104** 056801

[64] Todd K, Chou H-T, Amasha S and Goldhaber-Gordon D 2009 Quantum dot behavior in graphene nanoconstrictions *Nano Lett.* **9** 416–21

[65] Tien D H, Park J-Y, Kim K B, Lee N, Choi T, Kim P, Taniguchi T, Watanabe K and Seo Y 2016 Study of graphene-based 2D-heterostructure device fabricated by all-dry transfer process *ACS Appl. Mater. Interfaces* **8** 3072–78

[66] Kretinin A V *et al* 2014 Electronic properties of graphene encapsulated with different two-dimensional atomic crystals *Nano Lett.* **14** 3270–76

[67] Molitor F, Güttinger J, Stampfer C, Graf D, Ihn T and Ensslin K 2007 Local gating of a graphene Hall bar by graphene side gates *Phys. Rev.* B **76** 245426

[68] Wu S *et al* 2018 Magnetotransport properties of graphene nanoribbons with zigzag edges *Phys. Rev. Lett.* **120** 216601

[69] Kalhor N, Boden S A and Mizuta H 2014 Sub-10 nm patterning by focused He-ion beam milling for fabrication of downscaled graphene nano devices *Microelectron. Eng.* **114** 70–7

[70] Nanda G, Hlawacek G, Goswami S, Watanabe K, Taniguchi T and Alkemade P F A 2017 Electronic transport in helium-ion-beam etched encapsulated graphene nanoribbons *Carbon* **119** 419–25

[71] Oliveira M H Jr, Lopes J M J, Schumann T, Galves L A, Ramsteiner M, Berlin K, Trampert A and Riechert H 2015 Synthesis of quasi-free-standing bilayer graphene nanoribbons on SiC surfaces *Nat. Commun.* **6** 7632

[72] Miccoli I, Aprojanz J, Baringhaus J, Lichtenstein T, Galves L A, Lopes J M J and Tegenkamp C 2017 Quasi-free-standing bilayer graphene nanoribbons probed by electronic transport *Appl. Phys. Lett.* **110** 051601

[73] Emtsev K V, Zakharov A A, Coletti C, Forti S and Starke U 2011 Ambipolar doping in quasifree epitaxial graphene on SiC(0001) controlled by Ge intercalation *Phys. Rev.* B **84** 125423

[74] Baringhaus J, Sthr A, Forti S, Starke U and Tegenkamp C 2015 Ballistic bipolar junctions in chemically gated graphene ribbons *Sci. Rep.* **5** 9955

[75] Zakharov A A, Vinogradov N A, Aprojanz J, Nguyen T T N, Tegenkamp C, Struzzi C, Iakimov T, Yakimova R and Jokubavicius V 2019 Wafer scale growth and characterization of edge specific graphene nanoribbons for nanoelectronics *ACS Appl. Nano Mater.* **2** 156–62

[76] Oliveira M H *et al* 2013 Formation of high-quality quasi-free-standing bilayer graphene on SiC(0001) by oxygen intercalation upon annealing in air *Carbon* **52** 83–9

[77] Zhu W, Perebeinos V, Freitag M and Avouris P 2009 Carrier scattering, mobilities, and electrostatic potential in monolayer, bilayer, and trilayer graphene *Phys. Rev.* B **80** 235402

[78] Bolotin K I, Sikes K J, Hone J, Stormer H L and Kim P 2008 Temperature-dependent transport in suspended graphene *Phys. Rev. Lett.* **101** 096802

[79] Baringhaus J, Edler F, Neumann C, Stampfer C, Forti S, Starke U and Tegenkamp C 2013 Local transport measurements on epitaxial graphene *Appl. Phys. Lett.* **103** 111604

[80] Fang X-Y, Yu X-X, Zheng H-M, Jin H-B, Wang L and Cao M-S 2015 Temperature- and thickness-dependent electrical conductivity of few-layer graphene and graphene nanosheets *Phys. Lett.* A **379** 2245–51

[81] Povilas P and Antanas K 2012 Variable range hopping and/or phonon-assisted tunneling mechanism of electronic transport in polymers and carbon nanotubes *Open Phys.* **10** 271

[82] Edman L, Sundqvist B, McRae E and Litvin-Staszewska E 1998 Electrical resistivity of single-crystal graphite under pressure: an anisotropic three-dimensional semimetal *Phys. Rev.* B **57** 6227–30

[83] Ferrari A C and Basko D M 2013 Raman spectroscopy as a versatile tool for studying the properties of graphene *Nat. Nanotechnol.* **8** 235

[84] Katsnelson M I, Novoselov K S and Geim A K 2006 Chiral tunnelling and the Klein paradox in graphene *Nat. Phys.* **2** 620

[85] Wilmart Q, Berrada S, Torrin D, Nguyen V H, Fve G, Berroir J-M, Dollfus P and Plaais B 2014 A Klein-tunneling transistor with ballistic graphene *2D Mater.* **1** 011006

[86] Cheianov V V, Fal'ko V and Altshuler B L 2007 The focusing of electron flow and a veselago lens in graphene p-n junctions *Science* **315** 1252–55

[87] Rickhaus P, Maurand R, Liu M-H, Weiss M, Richter K and Schnenberger C 2013 Ballistic interferences in suspended graphene *Nat. Commun.* **4** 2342

[88] Gorbachev R V, Mayorov A S, Savchenko A K, Horsell D W and Guinea F 2008 Conductance of p-n-p graphene structures with air-bridge top gates *Nano Lett.* **8** 1995–99

[89] Baringhaus J, Sthr A, Forti S, Krasnikov S A, Zakharov A A, Starke U and Tegenkamp C 2014 Bipolar gating of epitaxial graphene by intercalation of Ge *Appl. Phys. Lett.* **104** 261602

[90] Kim H, Dugerjav O, Lkhagvasuren A and Seo J M 2016 Origin of ambipolar graphene doping induced by the ordered Ge film intercalated on SiC(0001) *Carbon* **108** 154–64

[91] Emtsev K V *et al* 2009 Towards wafer-size graphene layers by atmospheric pressure graphitization of silicon carbide *Nat. Mater.* **8** 203

[92] Kruskopf M *et al* 2016 Comeback of epitaxial graphene for electronics: large-area growth of bilayer-free graphene on SiC *2D Mater.* **3** 041002

[93] Baringhaus J, Edler F and Tegenkamp C 2013 Edge-states in graphene nanoribbons: a combined spectroscopy and transport study *J. Phys.: Condens. Matter.* **25** 392001
[94] Allain P E and Fuchs J N 2011 Klein tunneling in graphene: optics with massless electrons *Europhys. Phys. J.* B **83** 301
[95] Cheianov V V and Fal'ko V I 2006 Selective transmission of Dirac electrons and ballistic magnetoresistance of *n-p* junctions in graphene *Phys. Rev.* B **74** 041403
[96] Nevius M S, Wang F, Mathieu C, Barrett N, Sala A, Menteş T O, Locatelli A and Conrad E H 2014 The bottom-up growth of edge specific graphene nanoribbons *Nano Lett.* **14** 6080–86
[97] Miettinen A L *et al* 2019 Edge states and ballistic transport in zig-zag graphene ribbons: the role of SiC polytypes arXiv:1903.05185
[98] Kajiwara T, Nakamori Y, Visikovskiy A, Iimori T, Komori F, Nakatsuji K, Mase K and Tanaka S 2013 Graphene nanoribbons on vicinal SiC surfaces by molecular beam epitaxy *Phys. Rev.* B **87** 121407
[99] Ienaga K *et al* 2017 Modulation of electron-phonon coupling in one-dimensionally nanorippled graphene on a macrofacet of 6h-SiC *Nano Lett.* **17** 3527–32
[100] Aprojanz J, Power S R, Bampoulis P, Roche S, Jauho A-P, Zandvliet H J W, Zakharov A A and Tegenkamp C 2018 Ballistic tracks in graphene nanoribbons *Nat. Commun.* **9** 4426
[101] Baringhaus J, Aprojanz J, Wiegand J, Laube D, Halbauer M, Hbner J, Oestreich M and Tegenkamp C 2015 Growth and characterization of sidewall graphene nanoribbons *Appl. Phys. Lett.* **106** 043109
[102] Baringhaus J, Settnes M, Aprojanz J, Power S R, Jauho A-P and Tegenkamp C 2016 Electron interference in ballistic graphene nanoconstrictions *Phys. Rev. Lett.* **116** 186602
[103] Zhang Y, Brar V W, Wang F, Girit C, Yayon Y, Panlasigui M, Zettl A and Crommie M F 2008 Giant phonon-induced conductance in scanning tunnelling spectroscopy of gate-tunable graphene *Nat. Phys.* **4** 627
[104] Wehling T O, Grigorenko I, Lichtenstein A I and Balatsky A V 2008 Phonon-mediated tunneling into graphene *Phys. Rev. Lett.* **101** 216803
[105] Aprojanz J, Bampoulis P, Zakharov A A, Zandvliet H J W and Tegenkamp C 2019 Nanoscale imaging of electric pathways in epitaxial graphene nanoribbons *Nano Res.* **12** 1697–702
[106] Berger C *et al* 2006 Electronic confinement and coherence in patterned epitaxial graphene *Science* **312** 1191–96
[107] Büttiker M 1986 Four-terminal phase-coherent conductance *Phys. Rev. Lett.* **57** 1761–64
[108] Shi X, Boden S A, Robinson A and Lawson R 2016 *Scanning Helium Ion Beam Lithography in Frontiers of Nanoscience* vol 11 (Amsterdam: Elsevier), ch 17 pp 563–94

Chapter 7

Quantum transport in graphene nanoribbons in the presence of disorder

Alejandro Lopez-Bezanilla, Alessandro Cresti, Blanca Biel, Jean-Christophe Charlier and Stephan Roche

7.1 Introduction

Carbon nanotubes (CNTs) [1], two-dimensional graphene [2] and graphene nanoribbons (GNRs) [3] offer true possibilities for designing efficient carbon-based field-effect transistors as discussed in other chapters of this book (see also [4]). These materials show an exceptional ability to transport charge carriers and display some of the most exotic features of quantum transport. The origin of these properties resides in the linear band dispersion of low-energy electrons, and in the presence of a pseudospin degree of freedom. Around the charge neutrality point, electrons behave as massless Dirac fermions, which leads to unique quantum phenomena such as Klein tunneling [5] and weak antilocalization (WAL) [6, 7]. These fascinating properties of clean graphene-based systems can be further tuned and diversified by chemical modifications of the underlying π-conjugated network [8].

Both experimental and theoretical studies have traditionally considered external modifications like adsorbed atoms and molecules, substitutional impurities and functional groups as genuine ways to enhance material and device functionalities. For example, substitution of carbon atoms by boron or nitrogen atoms produces a doping effect, with the advantage of preserving the original geometry of the graphene lattice. As discussed hereafter, this type of doping induces intrinsic electron–hole transport asymmetry and mobility gaps in GNRs. Differently, chemical modifications of GNRs through covalent functionalization produces large geometrical and electronic structure distortions. Such alterations of the graphene conducting properties constitute a chemical way to tune the charge carriers flow, thus representing a tremendous tool for engineering field-effect transistor devices. Within this backdrop, several issues need to be addressed, with the search of an efficient and weakly invasive chemical grafting process remaining a great challenge.

doi:10.1088/978-0-7503-1701-6ch7

© IOP Publishing Ltd 2020

Numerical simulations with first-principles methods based on density functional theory (DFT) furnish a realistic description of the structural as well as electronic properties of chemically complex low-dimensional materials. DFT computational modeling allows us to understand the chemistry of carbon-based materials upon external modifications, which in turn provides insight into their impact on both electronic and transport properties. However, the computational cost becomes prohibitive when considering several thousands of atoms, as in the case of defective systems with realistic size. To keep the accuracy of first-principles approaches and treat complex carbon-based materials in a realistic manner, new computational strategies must be developed.

Here, a strategy will be presented for quantum transport analysis that relies on bridging computationally efficient tight-binding (TB) techniques with first-principles calculations. The method has the advantage of system upscalability while capturing the dominant multiple scattering effects of any particular material composed of millions of atoms in the presence of defects, which may range from randomly distributed impurities to geometrically modified structures. This will be illustrated in section 7.5 for the case of substitutional boron doping and chemical attachment of external functional groups. A second approach consists in performing DFT simulations to obtain optimized nanoscale structures (typically a few hundred atoms), and in building larger size systems by matching techniques and by using the DFT-derived Hamiltonians as building blocks. This method becomes essential in the case of strong chemical modifications of otherwise clean π-conjugated material, as for methyl groups (CH_3) covalently attached to the surface of a nanoribbon.

In this chapter, after presenting DFT-based quantum transport methodologies and recapping the basics of the electronic properties of graphene, the mesoscopic properties of armchair GNRs in the presence of random distribution of defects, functional groups and doping impurities will be explored. The main concern will be to provide quantitative estimations of typical transport scale lengths (elastic mean free path and localization length) and related transport regimes in graphene nanoribbons.

7.2 Methods for electronic structure calculations

This section summarizes the main features of two widely employed solid state methods for the calculation of electronic structures in condensed matter, namely the TB method and DFT.

7.2.1 The TB method and its limitations

The TB method is an empirical approach that employs a minimal basis set composed of localized orbitals. Unlike first-principles DFT-based approaches, TB Hamiltonians are empirically determined by fitting the results obtained from experimental data or first-principles calculations. Whereas DFT results are more accurate than TB methods, DFT-based calculations need to be performed for each individual case, whereas the same TB model can be adapted to different geometries, strain, and environment conditions. TB models provide a good alternative to deal with large size systems and over long time scales that cannot currently be simulated by first-principles methods.

For systems such as graphene and graphene nanoribbons, the expansion of the Hamiltonian and electronic wavefunctions in terms of the valence orbitals of the constituting atoms provides a reasonable description of the low-energy electronic features while keeping the computational cost at a minimum. The four valence orbitals of a carbon atom in a two-dimensional graphene sheet form three in-plane sp^2 hybridized orbitals and one plane-perpendicular p_z orbital. The low-energy σ-bonds are constituted by the sp^2 orbitals, and they ensure the mechanical stability. The π-bonds are built through the combination of the p_z orbitals which lie closer to the Fermi level. The electrons populating p_z orbitals are thus responsible for the electronic transport at low energies around the charge neutrality point. Hence, for electron transport properties analysis, a single p_z orbital TB model with first-neighbor interactions can be adopted, which reproduces the low-energy band structure with reasonable accuracy. However, when sp^2 orbitals come into play, as in the case of covalent functionalization of graphene by external addends, the single-orbital TB description fails. In this case, self-consistent DFT-based calculations are mandatory to obtain reliable descriptions of the equilibrium atomic positions and electronic states of the system under study.

To sum up, while TB methods are computationally efficient, DFT-based calculations provide an accurate description of the material charge density, which is essential for a proper determination of equilibrium atomic positions and electronic states of a complex system.

7.2.2 Localized basis sets in DFT

The DFT is a powerful quantum-mechanical approach to access structural and electronic properties of materials that is independent of empirical data, although approximations in treating electronic interactions are made. Hereafter, the use of DFT-based simulations to derive sophisticated TB parameterizations will be explained. These 'advanced' TB models provide a sufficiently accurate description of the transport properties while minimizing the computational cost.

DFT powerfulness relies on considering the system of interacting electrons equivalent to one of non-interacting electrons under the action of an external effective potential, with the requirement that the non-interacting system has the same ground-state electron density as the interacting one. This convenient 'ansatz' allows us to work with independent-particle equations, from which single-particle wavefunctions can be obtained to build the electron density for the interacting system. Within the Kohn–Sham approach (KS), the many-body Hamiltonian can be separated into a 'known' independent-particle functional and an 'unknown' exchange-correlation functional. Once the exchange-correlation functional has been chosen, the basis functions used to solve the set of KS equations have to be defined. Basis functions of various types are commonly used to expand the KS eigenstates, each of them presenting specific advantages and drawbacks.

As presented below in section 7.3, to calculate transport properties within the Landauer–Büttiker (LB) formalism using DFT Hamiltonians, the computation of the lattice Green's function must be done on a localized orbital basis. Several solutions have been suggested in order to extract a localized representation suitable for the LB formalism by projecting the quantum-mechanical eigenstate onto a localized atomic basis set. An interesting approach to the calculation of coherent transport properties of nanostructures from first-principles consists in the use of maximally localized Wannier functions [9]. This methodology has been successfully developed and implemented by Lee and Marzari for the study of carbon nanotube transport properties [10, 11]. Alternatively, solving the Kohn–Sham DFT problem by expressing the eigenstates as a linear combination of localized atomic-like orbitals (LCAO) straightforwardly leads to a sparse Hamiltonian provided that a local or semi-local exchange-correlation functional is used. These conceptual and technical requirements are implemented in the fully self-consistent SIESTA code [12, 13], which is based on a flexible and strictly confined linear combination of atomic-like orbitals [14]. The KS orbitals are thus expanded into localized orbitals that allow one to cope with some of the order-N method requirements for fast computation using minimal basis sets. The numerical solution of the KS equations is based on a linear expansion of the wave functions in terms of an atomic-like basis set:

$$\Psi_n(\mathbf{r}) = \sum_m \mathrm{c}_{nm}\phi_m(\mathbf{r}), \qquad (7.1)$$

where $\{\phi_m(\mathbf{r})\}$ is a set of orbital wave functions centered on the atoms. The main advantage of atomic orbitals is their efficiency and the main disadvantage is the lack of systematicity to improve the basis set, a limitation that is partially overcome by a careful choice of the parameters defining the basis functions shape until an adequate convergence is reached.

7.3 The LB quantum transport model

This section provides a brief presentation of the LB transport methodology as well as its use within a first-principles electronic structure framework. The analysis of the transport properties of disordered GNRs is split into the following steps. First, a study the perturbation effect that single and isolated sources of disorder (dopants, edge defects, grafted molecules) have on the geometry and on the electronic properties of an otherwise ideal GNR is performed. Afterwards, a study of the transport properties of more realistic GNRs is carried out by calculating the conductance of long and non-periodic GNRs and performing a statistical average over a large number of different configurations of randomly distributed defects. This approach combines first-principles calculations to obtain the effective one-electron Hamiltonian for the defected GNRs with Green's function techniques framed within the LB formalism, thus enabling the calculation of the conductance for very long (up to the micron scale) systems within the DFT accuracy.

The LB formalism is particularly suited for studying electron transport in phase-coherent systems composed of a scattering region in between perfect leads. The key idea of the LB formalism is to express the electric current flowing through a certain system in terms of the probability of an electron to be transmitted across the system itself. The device is typically constituted of a channel (the scattering region) connected to two semi-infinite leads at equilibrium with ideal electron reservoirs; see figure 7.1. The LB method is related to the transmission concept: the incoming electronic wavepackets propagating into the channel will scatter and interfere with a certain probability to be either reflected or transmitted through the device. Material impurities such as dopants, structural imperfections or edge roughness, induce electron scattering, thus limiting the electronic transmission probability [15].

Within the LB approach, the electronic conductance is:

$$G = \frac{2e^2}{h}\sum_{n=1}^{N} T_n, \tag{7.2}$$

where the sum runs over all the different conducting channels at the Fermi energy, T_n are their transmission coefficients, $\frac{e^2}{h}$ is the quantum of conductance and the factor 2 takes into account the spin degeneracy. This expression assumes that charge propagates coherently through the system, while all of the inelastic processes take place in the reservoirs, where electrons relax to thermodynamic equilibrium.

Since the early development of electronic transport simulations in semi-infinite crystals, efficient order-N Green's function techniques, such as the decimation techniques [16, 17], have been extensively employed and optimized. These methods rely on the expansion of Green's function matrices in terms of a localized basis set, and thus have been mainly implemented into semi-empirical TB models. By using compact Hamiltonians derived from DFT-based methods on localized basis sets, similar procedures can be implemented. The sparse nature of the Hamiltonian expressed on a localized-basis (atomic orbitals, Wannier functions, etc) allows the implementation of recursive techniques, which scale linearly with the length of the conducting channel. The computation of the conductance of clean and defected GNRs is performed using a multichannel LB technique as described in [18–21]. As explained in the previous section, the channel is defined between two semi-infinite right and left leads (see figure 7.1), which are in contact with electron reservoirs at fixed chemical potentials μ_1 and μ_2. The transmission coefficient $T(\epsilon)$ is then obtained by solving the quantum mechanical problem of a single electron scattered

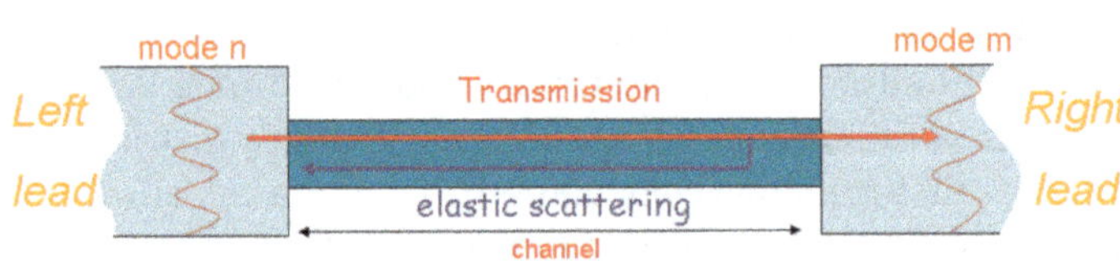

Figure 7.1. An electron with a given energy enters into the channel from the left lead. Within the Landauer's model, the incoming wavepacket in the n-mode has a certain probability to either reach the left lead or be elastically backscattered. This probability is related to the number of available transmission channels.

within the channel. At zero temperature and in the low bias limit ($\mu_1 - \mu_2 = eV \to 0$), the conductance writes $G = G_0 T(\epsilon)$, where $G_0 = 2e^2/h$ is the quantum conductance and ϵ denotes the energy of the incident charge carrier. Experimentally, the energy ϵ can be varied through a capacitive coupling between the ribbon channel and a gate electrode. The transmission is calculated by evaluating the retarded (advanced) Green's functions of the system

$$\mathcal{G}^{\pm}(\epsilon) = \left\{\epsilon S - H - \Sigma_L^{\pm}(\epsilon) - \Sigma_R^{\pm}(\epsilon)\right\}^{-1} \tag{7.3}$$

with $\Sigma_{L(R)}^{\pm}(\epsilon)$ the self-energies describing the coupling of the channel to the left (right) lead. These quantities are related to the transmission coefficient by the Fisher and Lee relationship [15]

$$T(\epsilon) = \mathrm{tr}\{\Gamma_L(\epsilon)\mathcal{G}^{+}(\epsilon)\Gamma_R(\epsilon)\mathcal{G}^{-}(\epsilon)\}, \tag{7.4}$$

where $\Gamma_{L(R)}(\epsilon) = i\{\Sigma_{L(R)}^{+}(\epsilon) - \Sigma_{L(R)}^{-}(\epsilon)\}$, and tr denotes the trace operator.

7.4 Mesoscopic DFT-based transport calculations of disordered nanoribbons

This section illustrates the use of DFT tools to explore the structural disruption of the sp^2 carbon atom orbital hybridization in armchair GNRs (aGNRs) and the corresponding effect on electronic transport properties. π-bonds rupture and additional σ-bond production, i.e. transition from sp^2 to sp^3 hybridization, is the main effect on the local charge distribution of some types of chemisorption on carbon atoms of graphene-based materials. The consequence on the material electronic structure is the formation of impurity scattering centers, which introduce important drawbacks for electronic transport.

7.4.1 Building blocks from DFT calculations

The use of DFT Hamiltonians within an LB approach to study transport properties at the mesoscopic scale is well-documented in the literature. Several groups have made extensive use of these techniques to analyze charge transport properties in disordered low-dimensional structures such as carbon nanotubes, silicon nanowires and graphene nanoribbons [11, 21–25]. The general approach is first to perform a set of DFT calculations to obtain the Hamiltonian matrices associated with short defected sections (for instance, sections of functionalized GNRs), which will constitute the building blocks (BBs) for creating long and randomly disordered ribbons. The coupling between BBs is assumed to be similar to the coupling between bulk layers of ideal GNRs. For this reason, the number of layers included in the supercell of the *ab initio* simulation must be large enough so that the surface layers of the supercell are not affected by the perturbation caused by the defect [26]. As for any TB-like model, this coupling or 'hopping' matrix must contain interactions only between atoms in adjacent BBs. The size of the Hamiltonian matrix of this defected BBs will determine the maximum size of the matrices involved in the renormalization process. The total disordered system is then built up by randomly assembling such elementary parts containing a single or a few defects with sections of pristine (clean) nanoribbons. The leads are simulated as semi-infinite perfect GNRs, built up from short sections of ideal GNRs. This allows us to study transport through very long GNRs by using standard decimation techniques for the calculation of Green's functions and transmission probabilities [22]. All first-principles calculations reported in this chapter were performed using a localized atomic-like basis set in the local density approximation (LDA) (SIESTA code [12, 13]), with a double-ζ basis for each atom [23]. Spin polarized calculations within LSDA yielded no magnetic ground states, showing that the triplet state lays 240 meV above the singlet state. Structures were relaxed until residual forces were smaller than 0.01 eV/Å.

7.5 Boron and nitrogen substitutional doping

One of the most striking advances in GNRs synthesis has been the ability to deterministically place atomic dopants in selected, precise positions along and across the graphene ribbons. Chemical doping offers a route to tune the electronic properties of graphene through the local modification of potential energy and the incorporation of resonant scatters. The introduction of chemical dopants as a possibility to tune in a controlled way the GNRs electronic and transport properties was envisioned as early as 2008 [27] in Ruoff's famous article '*Call All Chemists*', and since then his predictions were supported by numerous simulations of the transport properties of both armchair and zigzag GNRs with various functionalizations [28–30]. However, in order to experimentally realize such theoretical predictions, it was essential not only to fabricate ribbons with atomically sharp edges, but also to control the deterministic placement of the substitutional dopants at the desired positions. Both requisites have since then been achieved thanks to the mastery of molecular engineering tools, such as bottom-up synthetic methods based on carefully selected molecular precursors with the desired doping species [31–39].

Among the different chemical species, boron and nitrogen are natural dopants for carbon materials. Numerous research efforts, both experimental and theoretical, in understanding the effect of B and N doping on 2D graphene and GNRs [29, 40, 41] have confirmed this method to be a successful way of tuning the properties of low-dimensional graphene-based materials. In our studies of boron and nitrogen doping, we have combined, as previously explained, the DFT results for the isolated dopant case with a fine parameterization of the TB model according to these first-principles results. To this aim, extensive simulations using the SIESTA code are first performed for an aGNR with one of its carbon atoms substituted by a boron [28] or nitrogen [28, 42] dopant (simulation details can be found in [28]). This self-consistent calculation provides the profile of the scattering potential around the impurity. In a second step, the onsite and hopping self-consistent Hamiltonian matrix elements (on a localized basis set) are used to build the TB Hamiltonian. This technique has been successfully employed in the study of electronic properties in disordered carbon nanotubes [42–44] and two-dimensional doped graphene [45].

7.5.1 Electronic and transport properties in single-doped ribbons

We will first focus on a 35-aGNR, i.e. an aGNR with 35 dimer lines across the ribbons width (≈4.4 nm). The nearest-neighbor TB model predicts a crossing of the $\pi - \pi^*$ bands at the point $3/4(\pi/a)$ of the Brillouin zone [46]. However, we have already mentioned that *ab initio* calculations [47, 48] predict all aGNRs to be semiconducting. The 35-aGNR belongs to the family with $N = 3p + 2$ and has thus a smaller gap than aGNRs in the other two families. The conductance profile for a single dopant turns out to be strongly dependent of the position of the impurity with respect to the ribbon edge [28], as shown in figure 7.2 (left). This feature results in a marked electron–hole conductance asymmetry, which leads to a significant suppression of the hole transport when doped regions of mesoscopic length are considered. In contrast, electron conductance remains very close to the conductance quantum, thus suggesting the robustness of the (quasi-)ballistic regime.

One interesting quantum phenomenon predicted in this work is the full suppression of backscattering on the $\pi - \pi^*$ plateau (black curve in figure 7.2 (left)), when the impurity preserves the mirror symmetry of armchair ribbons and hence the well-defined parity of the wavefunction, thus forbidding the scattering between even and odd states. This effect has been only recently observed by Carbonell-Sanroma *et al* [32] in GNRs doped with boron pairs. In this case, the first valence band of the pristine ribbon was confined while allowing an efficient electron transmission of the second one, due to the symmetry matching between the electronic wavefunctions of the states from the pristine nanoribbons and those localized at the boron pairs.

7.5.2 Mesoscopic transport properties in randomly-doped ribbons

To further explore the case of aGNRs with a more realistic width of ≈10 nm we consider the 80- and the 81-aGNRs. For these systems, a 0.2% doping rate is enough to achieve a mobility gap of ≈1 eV (see figure 7.2 (right)), while keeping high

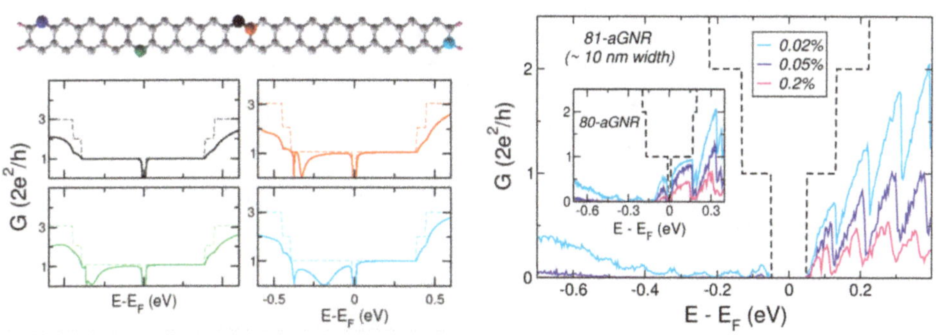

Figure 7.2. (Left) Conductance as a function of energy for the 35-aGNR with single B dopants at different positions across the ribbon width. The dopant position for each colored curve is shown in the unit cell plot of the GNR (top) by an atom with the same color. (Right) Main panel: average (over 500 disorder configurations) conductance as a function of energy for a 1 μm length semiconducting 81-aGNR at three selected doping rates (≈ 0.02%, 0.05% and 0.2%, from top to bottom). Inset: same as in the main frame for the pseudo-metallic 80-aGNR.

conductance values in the conduction band for both types of wider GNRs. In contrast to the situation of narrower ribbons, higher doping rates are required to achieve a full suppression of conductance in the valence band. This is due to the fact that backscattering in aGNRs is highly dependent on the dopant position, with a maximum impact when the dopant is close to (or right at) the ribbon edge [28]. As a consequence of the uniform random distribution of dopants, the probability to find an impurity close to the edges decreases with increasing ribbon width, and will be, for an 80-aGNR, almost one half that for a 35-aGNR, thus leading to a lower impact on the conductance for a similar doping rate (figure 7.3). An important issue concerns the sample-to-sample conductance variabillity driven by different configurations of impurities in low-dimensional systems. This issue has been investigated in [49] for boron-doped GNRs. As the ribbon width is reduced, larger fluctuations are obtained. This suggests the need for an optimization strategy in order to maximize the mobility gaps in each sample, while preserving a low variability level between different disorder realizations.

7.5.3 Topological defects

The issue of edge disorder has attracted much attention from the graphene community, see for instance [50–53]. Defects at the edges are expected to occur in the fabrication process of the ribbons and they might affect the electronic and transport properties drastically. It is thus of great concern to explore the impact of edge defects.

Most of the investigations on this issue have been performed by adopting simplified topological TB models or even more general Anderson-like disorder models [51, 52, 54]. However, simplified topological TB models might not provide a realistic description of the electronic structure for certain types of defects [55],

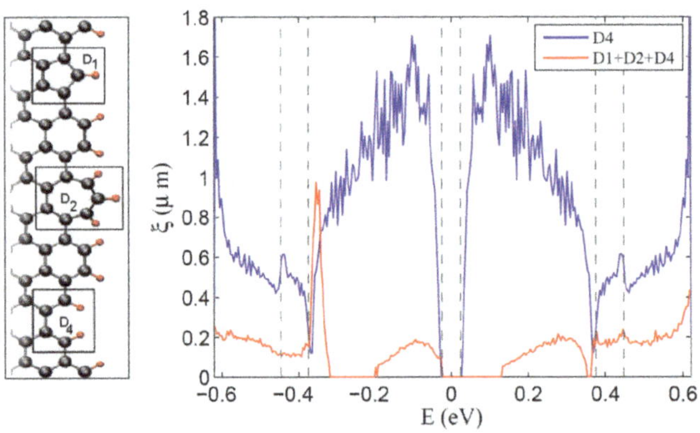

Figure 7.3. (Left) Three different types of edge defects considered in the simulation. (Right) Localization length ξ for a 35-aGNR with a 6×10^{-2} nm^{-1} density of D_4 defects (blue line) and a mix of $\{D_1, D_2, D_4\}$ defects. The vertical dashed lines correspond to the position of the van Hove singularities. In some energy regions, ξ is very short (below 100 nm) and it cannot be extracted from the data. In this case, we set $\xi = 0$. The localization length for the D_4 defects is up to almost one order of magnitude longer than in the case of mixed defects.

because they do not take into account charge transfer between atoms. A more accurate edge-disorder model is thus elaborated by tuning the nearest-neighbor TB Hamiltonian in order to reproduce the DFT-based conductance profile for a single defect in an infinite GNR [18, 55]. Mesoscopic transport calculations are then performed using this Hamiltonian. First-principles transport calculations [55] have shown how, depending on the actual geometry of the edge defect and its degree of hydrogenation, completely different transport mechanisms are observed. In the case of the monohydrogenated pentagon (heptagon) defects, an effective acceptor (donor) character results in a strong electron-hole conductance asymmetry. In contrast, weak backscattering is obtained for defects that preserve the benzenoid structure of graphene.

Thus, the transport profiles and the values for the mean free path ℓ_e and the localization length ξ are found to be very much dependent on the actual edge reconstruction [55]. To illustrate such a feature, we consider the presence of a random distribution of pentagonal (D_4) defects, and a mix of pentagonal (D_1), heptagonal (D_2) and 'missing hexagon' (D_4) defects for the same edge defect density. For both defect distributions, the conductance decay is found to be rather homogeneous within the first electron and hole plateaus. While the marked electron–hole asymmetries associated with the D_1 and D_2 defects compensate each other, the presence of odd membered rings continues to crucially impact the electron and hole localization. Defects that are non-compliant with the benzenoid character of aGNRs (i.e. D_1 and D_2) backscatter electrons more severely, even for the low defect concentration considered here. This is clearly seen by comparing the localization lengths (figure 7.4(d)) as a function of the carrier energy. While the distribution of D_4 defects gives rise to a localization length $\xi \sim 1$ μm for low energy carriers, the introduction of charged defects strongly reduces ξ in the [−0.5, 0.5] eV energy window. The fluctuations of ξ in correspondence of the van Hove singularities for D_4 defects are due to the increased scattering induced by the high density of

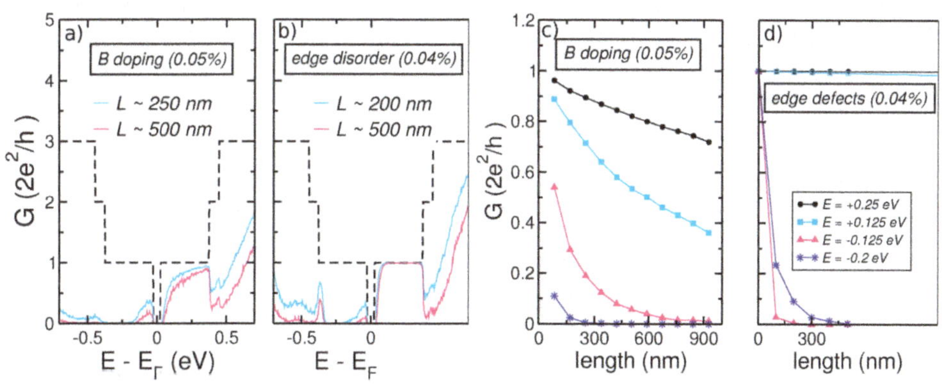

Figure 7.4. (a) Conductance as a function of the energy for the 35-aGNR at a B atom doping rate of 0.1%, and in (b) edge defect density of 0.04% for several fixed lengths. (c) Conductance as a function of the length for the 35-aGNR at a B atom doping rate of 0.05%, and in (d) edge defect density of 0.04% for different values of the energy *E*.

states at these points. In the case of the mixed disorder, these fluctuations are absent or considerably reduced owing to the strong smearing effect that disorder produces in the density of states.

We now pinpoint the main similarities and differences with respect to the case of chemical doping. Among the possible reconstructions [56], we have selected the pentagon-like edge defect, since its conductance profile reminds us closely of that of the boron impurity. Figure 7.4 shows the conductance profile as a function of energy for a 35-aGNR in the case of boron doping (a) and pentagon-like defects (b). The conductance curves average over more than 400 disorder configurations (i.e. different distributions of dopants or defects compatible with a given doping rate or defect density). As can be seen in the plot, both types of disorder lead to the onset of mobility gaps in the valence band of the aGNR, although the boron doping seems to be more efficient and opens up a larger gap for a similar length. This is a consequence of the wider distribution of quasibound states over the entire valence band in the first conductance plateau, due to the random distribution of dopants across the ribbon width and the strong dependence of the scattering potential on the dopant position with respect to the ribbon edge. It is however important to point out that the electron conductance, although keeping high values for both disorder models, is better preserved in the case of this particular edge disorder. The shape of the scattering potential, which leads to a very robust conductance in the first conduction band, is responsible for the very long mean free paths in the corresponding energy range. Indeed, as shown in figure 7.4(b), the averaged conductance at those energy values remains very close to its maximum value G_0.

At last, recent outbreaks in bottom-up chemical techniques have demonstrated the synthesis of atomically perfect zigzag graphene nanoribbons (zGNRs) [57] and of their corresponding analogs with phenyl-edge functionalization [58]. Indeed, as depicted in figure 7.5(a), the monomer used would at first glance yield a perfectly periodic 1D structure. However, during the cyclodehydrogenation process, the phenyl groups undergo ring closure, which can happen towards two directions, inducing either perfectly periodic or also possibly non-periodic nanoribbons. These phenyl defects are strictly localized along the edges and do not modify the global

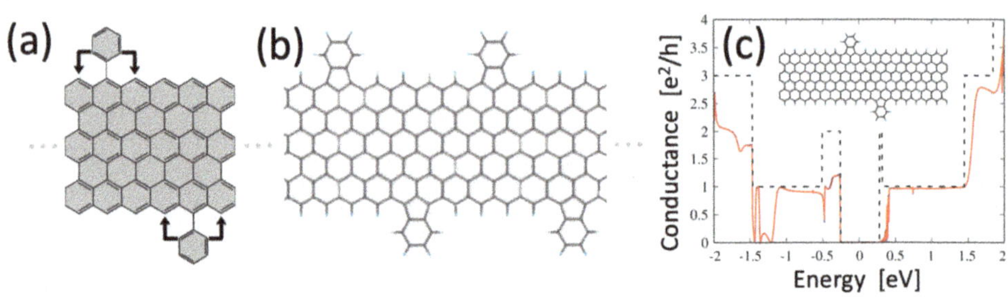

Figure 7.5. (a) Periodic 6-zGNRs with phenyl-edge defects before cyclodehydrogenation as obtained by Ruffieux *et al* [58]. This structure is not stable and during the cyclodehydrogenation process, the phenyl groups undergo ring close, falling either on one or the other side, possibly leading to ideal structures as in (b). (c) Conductance spectrum for 6-zGNRs with phenyl groups located on their edges, as illustrated in the inset.

structure of the ribbon (figure 7.5(b)). Since edge-localized states are responsible for the magnetic properties of zGNRs, the effect of such phenyl-edge-defect might lead to interesting spin-dependent transport properties. Using first-principle calculations and an LB approach, the spin-dependent electronic transmissions have been predicted for various phenyl-edge-modified zGNRs [59]. Figure 7.5(c) presents the conductance of a 6-zGNR with phenyl groups located at specific positions on both edges. Theoretical results suggest that the control of phenyl decoration at the zigzag edges could accurately tune the spin-polarized currents generated at the edges of the zGNRs [59]. Consequently, edge-engineering, with atomically controlled process, could open a new route to tailor the spin-dependent properties of these zGNRs, thus paving the way to the design of novel devices for future spintronics applications.

7.5.4 Covalent adsorption and sp^3-type defects

Covalent functionalization of graphene materials triggers the formation of saturated sp^3-bonds, which break the p_z-network symmetry [60, 61] and induce a severe disruption of the otherwise good conducting properties of clean graphene, even yielding the possibility to make graphene insulating in case of massive coverage [62]. DFT-based calculations are revealed as a powerful tool to elucidate the charge rearrangement due to the local rehybridization of carbon orbitals. After graphene discovery, several studies have been devoted to the hydrogenation of the graphite single layer to unravel its modified physical and chemical properties. In [63], the chemisorption of a single H atom was treated in detail, as well as the induced magnetic moment interaction as a consequence of grafting several atoms on the same sublattice. The creation of unpaired electrons leads in this case to a ferromagnetic graphene-addends interaction. Afterwards, it was shown that the most stable configuration is reached when the pair of atoms sits on different sublattices, which leads to a non-magnetic hybrid. This non-magnetic graphene state is lost again when a third atom H sits on a third C atom.

More aggressive grafting processes were studied by Sofo *et al* [64] by considering a graphene sheet completely covered by H atoms, distributed at both sides of the layer. The resulting diamond-like layer presents all the C atoms in sp^3-orbital hybridization instead of the original $sp^2 + p_z$ orbital conformation. This transformation of π- and π^*-orbitals to σ- and σ^*-orbitals results in a gap of $\sim$ 3 eV. A cohesive energy of 0.4 eV/(H atom) allows one to make the hydrogenation process reversible. In [65], the most stable configuration of one-side hydrogenated graphene was found when H atoms were bonded at first neighboring C atoms. Graphene is an aromatic material, meaning that it finds a minimum of configuration energy when π-electrons of carbon atoms are completely delocalized and all conjugated chemical bonds are equivalent. This symmetry disappears when a functional group is grafted onto graphene surface and breaks a π-bond, thus inducing an sp^2 to sp^3 rehybridization. In the bonding formation, a π-electron is used to establish the link graphene-addend and the second electron that participated in the $\pi-\pi$-bond remains unpaired in the vicinity of the new σ-bond. This electron is smeared in one of the sublattices and forms a magnetic moment. The spin density is localized $\sim$1nm around the graphene-addend bond, which is correlated with the crystal lattice distortion [66].

However, this situation is not stable and as soon as a second atom is chemisorbed on the graphene, the unpaired electron is used in the second sp^3 hybridization and the magnetic moment vanishes. Some studies reveal that the most stable configuration for grafting two hydrogen atoms is reached when they sit at first-neighbor carbon atoms but at different sides of the sheet, due to the minimal additional distortion upon second atom grafting. If only one side is available for the chemisorption, the most favorable configuration of two addends is the bonding with carbon atoms in opposite corners of the hexagon (1,4 or para-position) [67].

In the following, the case of methyl (CH_3) functional groups attached on GNRs is analyzed. A single group attached to graphene pulls out of the plane the C atoms to which it is attached, thus creating a C-CH_3 bond of 1.55 Å that is perpendicular to the plane, and an unpaired electron in its proximity. This unstable configuration lowers in energy when a second methyl group attatches an opposite carbon atom in a hexagonal graphene ring. This configuration of sp^3 defects by pairs quenches the radical and is always energetically preferred. Following prior studies for phenyl pairs [68] and OH/H functional groups [62], the grafted pair is such that the corresponding anchoring C-atoms are third nearest neighbors. It can also be observed that the length of the C-C bond near the edges is slightly shortened with respect to the central region of the ribbon.

Figure 7.6 shows the effect of a single pair of methyl groups at different positions across the nanoribbon width on the conductance of a 14-aGNR. Two zero conductance dips at energies above and below the charge neutrality point manifest a clear sp^3 signature of the single defect (as confirmed by the very similar pattern obtained for a single phenyl pair [68]). An electron–hole transport symmetry is

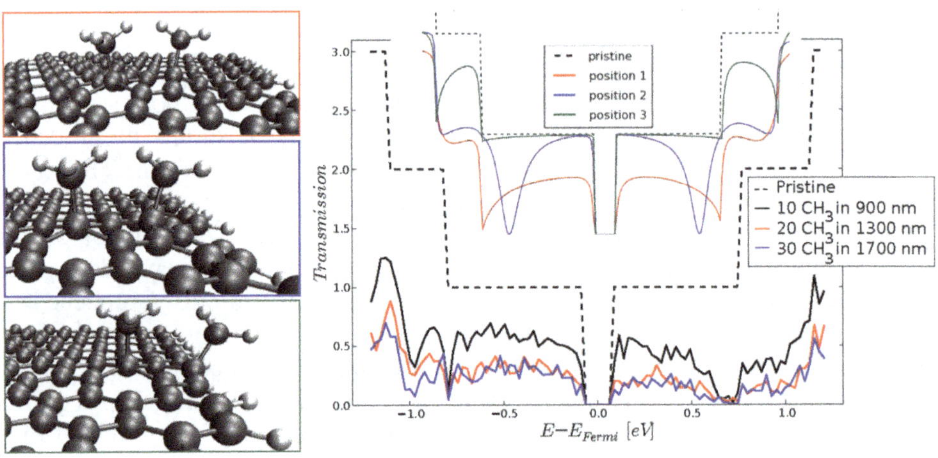

Figure 7.6. (Left) Three different configurations of methyl groups across a 14-aGNR. Inset of the right frame shows the conductance of the three configurations. (Right) The averaged conductance of the 14-aGNR for different number of grafted methyl groups and ribbon lengths. Quantized conductance of the pristine aGNR is shown with dotted lines.

observed for every configuration as a result of the symmetric charge distribution around the new bonds between the grafted groups and the GNR substrate. The symmetric backscattering profiles are also maintained at higher energies. The fact that the calculations for a grafted methyl pair show no direct doping to the graphene material supports the interpretation of [69] for a contact-induced doping effect not driven by the functionalization.

The conduction regime in long and disordered chemically functionalized aGNRs is analyzed by averaging the conductance values for ribbons with varying defect density, length and width. Randomness introduced in the defect distribution enhances backscattering as a result of quantum interferences between sp^3 defects, which ultimately lead to strong localization of states and full suppression of conductance. The results obtained for one single defect are amplified when multiple groups are attached to the ribbon, with a decrease of the average conductance when increasing the defect density from 10 (900 nm long ribbon) to 30 (1700 nm long ribbon) groups. A similar robustness of the conductance in the first plateau is seen when compared with higher energy subbands, similarly to the single defect case, but more pronounced for long disordered systems.

7.5.5 Bilayer graphene nanoribbons

In the previous paragraphs, it has been demonstrated numerically that the effective band gap of graphene can be enlarged by inducing backscattering through chemical modification of the lattice. Considering that graphene is often synthesized in bilayer form, it is interesting to analyze the impact that one impurity on one layer can have on the second layer. Furthermore, it is worth unveiling the effect on the layer lying on a substrate upon chemical modification of the layer exposed to the environment.

First-principles calculations followed by quantum transport analysis demonstrated that surface modification of one graphene layer induces inter-ribbon coupling beyond the van der Waals interaction due to the mixing of the electronic states of the parallel networks [70]. Figure 7.7 shows that typical conduction profiles obtained in monolayer GNRs upon modification with various types of impurities are diluted in a homogeneous type of conductance with a marked p-type-like doping

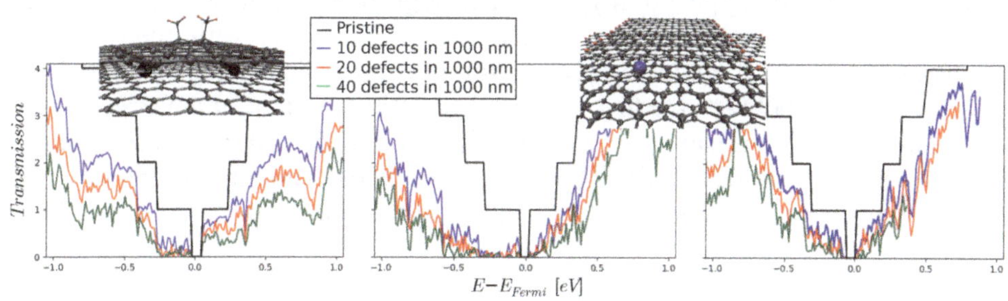

Figure 7.7. Averaged Landauer–Büttiker transmission coefficients for 1000 nm long bilayer graphene ribbons composed of 14-aGNRs chemically modified with methyl groups (left), and B and N atoms (middle and right, respectively) laid over a 30-aGNR. Quantized conductance of the pristine bilayer aGNR is shown with stepwise lines as a reference. Curves have been averaged over ten different random defective configurations.

character. The conductance of the chemically modified system is nearly independent of the external functional group or atom, and is characterized by electron–hole conduction asymmetries over a large energy range. The larger amount of conducting channels of the bilayer systems allows the conductance of the defected systems to remain more robust than in monolayered graphene for a larger number of defects.

Figure 7.7 shows that in the first and second plateaus the backscattering strength of the pairs of methyl groups, B, and N atoms is rather similar. The accumulation of up to 30 CH_3 groups and B atoms in the top ribbon enlarges the transport gap up to 0.25 eV in the hole band and reduces to less than one conducting channel the transport capability of the bilayer in the electron band in an energy range of up to 0.5 eV. Similar although more resistant to backscattering is the conductance profile observed for substitutional N atom doping, which almost drops to zero in the first plateau of the hole band for the same amount of defects.

For two parallel ribbons separated by typical van der Waals graphitic distance, chemical modification of one of the layers has an impact on the transport properties of both layers. This can be ascribed to the hybridization of their electronic states. Depending on the energy of the charge carriers and the degree of layer modification, several transport regimes are reached, ranging from the diffusive to the strongly localized, as a result of a strong electron–hole asymmetry induced by the external atoms and chemical groups. A common transport feature obtained for all the defects considered is the development of mobility gaps which are nearly independent of the type of structural or chemical modification induced in the bilayer.

7.6 Magnetotransport in disordered ribbons

Magnetotransport in graphene has attracted much interest from both fundamental and applicative viewpoints. At low magnetic fields, the peculiar pseudospin degree of freedom entails a transition from weak to anti-weak localization [6], which has turned out to be a powerful characterization tool for the estimation of the coherence, inter-valley scattering and intra-valley scattering times. At high fields, the Dirac-like nature of graphene gives rise to Landau levels (LLs) [71] with energy $E_n(B) \approx \pm 35\sqrt{B[T]}\sqrt{n}$ meV, where $n = 0, 1, 2....$ The resulting semi-integer quantum Hall effect [72] finds application in the metrological electrical resistance standard [73] as a possible replacement of GaAs-based Hall bars, with the advantage of operating at higher temperature and for larger currents thanks to the larger energy separation between the first LLs. In this section, we first summarize the main aspects of the quantum Hall effect in graphene, then we consider the case of narrow and ultranarrow ribbons under high magnetic fields, and finally analyze some sources of disorder that can affect the quality of the quantum Hall effect in larger ribbons.

7.6.1 Quantum Hall effect in graphene: Landau levels and edge states

The energy of the LLs in graphene can be straightforwardly obtained by considering the low-energy Dirac equation and by operating the minimal substitution $\pi = \mathbf{p} + e\mathbf{A}(\mathbf{r})/c$, where $\mathbf{A}(\mathbf{r}) = (-By, 0, 0)$ is the vector potential for a homogeneous and orthogonal magnetic field B in the first Landau gauge. The resulting Hamiltonian is

$$H = v_F \begin{pmatrix} 0 & \pi_x - i\pi_y \\ \pi_x + i\pi_y & 0 \end{pmatrix} = \frac{\hbar v_F}{\sqrt{2}\ell} \begin{pmatrix} 0 & \eta \\ \eta^\dagger & 0 \end{pmatrix}, \tag{7.5}$$

where $\ell = \sqrt{\hbar c/(eB)}$ is the magnetic length, and $\eta \equiv \ell/(\sqrt{2}\hbar)(\pi_x + i\pi_y)$ and $\eta^\dagger \equiv \ell/(\sqrt{2}\hbar)(\pi_x - i\pi_y)$ are creation and annihilation operators such that $[\eta, \eta^\dagger] = 1$. By squaring the Hamiltonian, we obtain the diagonal matrix

$$H^2 = \left(\frac{\hbar v_F}{\sqrt{2}\ell}\right)^2 \begin{pmatrix} \eta^\dagger\eta & 0 \\ 0 & \eta^\dagger\eta + 1 \end{pmatrix}, \tag{7.6}$$

where we identify the number operator $n = \eta^\dagger\eta$. The eigenvalues of H (i.e. the LLs) are thus given by $E_n = \pm\sqrt{e\hbar/(2c)}\, v_F \sqrt{B}\sqrt{n}$. Analogously to ordinary 2D electron gases [74], graphene under magnetic field is a topological (Chern) insulator [75], which gives origin to the quantum Hall effect, with the peculiarity of a semi-integer Hall conductivity quantization $\sigma_{xy} = 4e/h(n + 1/2)$ due to the presence of the $n = 0$ LL.

When laterally confining the system into a ribbon, such a topological nature entails the formation of metallic edge states. We verify this prediction by calculating the band structure of a 100 nm wide armchair ribbon described by the usual TB Hamiltonian with the inclusion of the Peierls phase factors [76] to take into account the magnetic field. Figures 7.8(a) and (b) report the band structure in the absence and in the presence of a 10 T magnetic field. We observe the formation of a double degenerate (including spin) LL with $n = 0$ and four-fold degenerate positive and negative LLs with $|n| > 0$. The development of the LLs requires that the magnetic length ℓ is smaller than the ribbon width. At the edges of the Brillouin zone, the energy of the bands increases. For example, at an energy $E = 50$ meV between the $n = 0$ LL and the $n = 1$ LL we have two states, one at positive and the other at negative wave number k; see the dots in figure 7.8(b). The corresponding probability density for these states is reported in figure 7.8(c) as a function of the position y along the ribbon section. The state with negative k is located at the bottom edge, while the state with positive k is located at the top edge. Moreover, they have opposite group velocities as a result of the opposite sign of the derivative of the

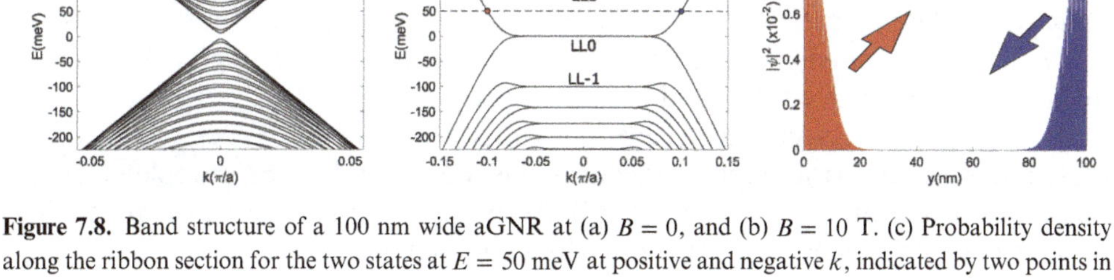

Figure 7.8. Band structure of a 100 nm wide aGNR at (a) $B = 0$, and (b) $B = 10$ T. (c) Probability density along the ribbon section for the two states at $E = 50$ meV at positive and negative k, indicated by two points in (a). The group velocities of the two states, indicated by the arrows, are opposite.

energy bands with respect to k. As a consequence, we end up with chiral edge states, i.e. edge states that bring electrons in opposite directions and separated by an insulating region. If this region is large enough and the two channels do not get into contact, backscattering is suppressed and the Hall conductance is perfectly and robustly quantized. Note that for holes ($E < 0$) the spatial chirality is reversed, meaning that the group velocities are opposite than for electrons.

7.6.2 Magnetotransport in disordered narrow and ultranarrow graphene ribbons

In ultranarrow graphene ribbons, due to the comparable size between ribbon width and magnetic length, we observe a competition between quantum and magnetic confinements [77, 78]. This results in the difficult formation of the LLs and, even in the case of well-developed edge states, in the fragility of their spatial chirality due to the narrow insulating region between opposite edge channels.

Poumirol *et al* [79] investigated magnetotransport in an ultrasmooth 11 nm wide ribbon obtained by chemically procedure [80]. Despite the high quality of the sample and the high magnetic field (up to 55 T), no conductance quantization is observed. However, a significant positive magnetoconductance is reported, which is clear indication of the partial formation of edge states at increasing magnetic field; see figure 7.9(a). To better understand this behavior, we performed quantum transport simulations under different conditions of disorder, and namely in the presence of edge roughness, short-range disorder and long-range disorder; see figures 7.9(b)–(d). From the qualitative comparison of the experimental and simulated results, it turns out that the experimental sample is most likely affected by a long-range disorder, due to charged impurities in the SiO_2 substrate, and a moderate edge roughness. The incomplete and progressive formation of edge states when increasing the magnetic field is clearly visible in the simulated spatial distribution of currents in figure 7.9(e), and is at the origin of the positive magnetoconductance. A strong contribution of short-range disorder is to exclude, because, in the investigated ultranarrow ribbon, it would lead to the complete quenching of the positive magnetoconductance, as seen in figure 7.9(b), due to the formation of circulating bulk states.

In contrast, conductance quantization was observed in slightly-larger ultra-smooth bilayer ribbons obtained by unzipping multiwall carbon nanotubes [81]. For a ribbon width around 20 nm, the conductance was found to quantize for magnetic fields above 25 T, which well corresponds to magnetic lengths shorter than the ribbon width itself. Remarkably enough, the possible additional edge states, which are intrinsic to the ribbon sections with zigzag geometry [82] and are not induced by the magnetic field, were found not to contribute to transport. Indeed, these states are non-chiral and extremely concentrated at the edges. As a consequence, a short-range edge disorder, as roughness, completely localizes them.

In larger monolayer graphene ribbons (with a width above 70 nm), the formation of the Landau levels at high magnetic fields was found to be complete, with a quantized two-terminal conductance [83]. Even more so, the edge states are sensitive to the ribbon edge geometry, which results in the signature of the valley degeneracy lifting in the magnetoresistance.

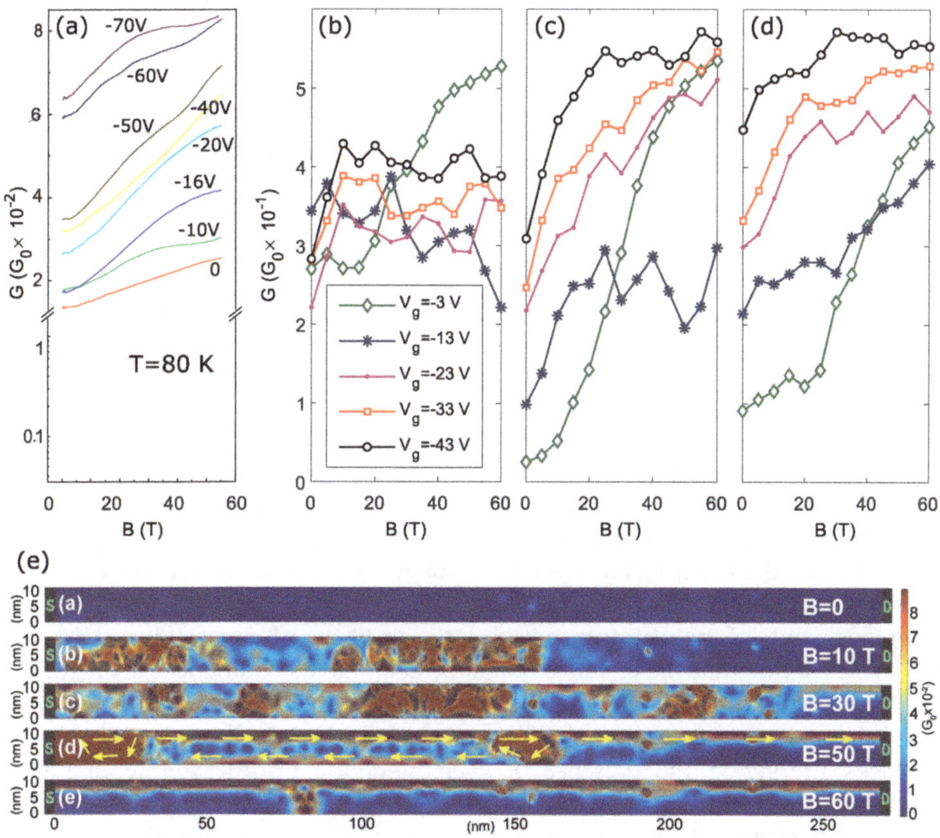

Figure 7.9. (a) Experimental magnetoconductance at 80 K for an 11 nm wide ribbon at different backgate potential V_g. Simulated zero-temperature differential magnetoconductance in an 11 nm wide armchair ribbon at different back gate potential V_g in the presence of (b) short-range Anderson disorder, (c) edge roughness, and (d) long-range Gaussian impurities. The disorder strength is calibrated to give a conductance in the same order as the experimental case. (e) Spectral current distribution at the electron energy $E = 200$ meV under different magnetic fields and in the presence of Gaussian disorder only. Electrons are injected from the left contact. The arrows in the penultimate panel indicate the path of the electrons, which are injected from the left along the partially formed bottom edge states and partly backscattered along the bottom edge due to the presence of defects. Reprinted with permission from [79], Copyright (2010) by the American Physical Society.

7.6.3 Spatial chirality breakdown in disordered large graphene ribbons

One of the main reasons for the breakdown of the spatial chirality in graphene ribbons is the presence of strong short-range disorder, as vacancies or adatoms, which generates local *inner*-edge states [84] able to induce, at high impurity density, a percolation path through the bulk [85–87]. To be more specific, the spatial extension of the states around the impurities is roughly comparable to the magnetic length ℓ. When the average distance between couples of neighbor impurities is shorter than ℓ, then the impurity states couple and form extended bulk states at inter-Landau-level energies. This result clearly indicates short-range disorder as a possible cause for dissipative quantum Hall effect. Resorting to high-quality graphene is thus central.

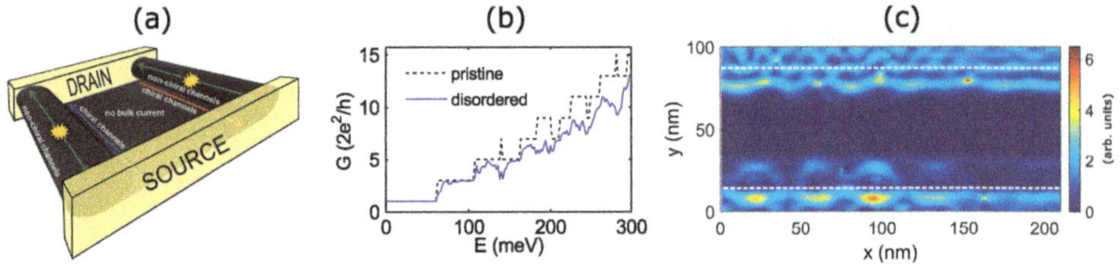

Figure 7.10. (a) Sketch of a ribbon with scrolled edges with indication of the non-chiral scroll channels and the chiral edge channels at the interface between the scrolls and the flat region of the ribbon. (b) Zero-temperature differential conductance as a function of the electron energy for a 100 nm wide ribbon with 20 nm wide scrolls (with two turns each) at $B = 20$ T, in the absence and in the presence of disorder. (c) Spectral transport current distribution at an electron energy $E = 170$ meV in the presence of disorder. The dashed lines indicate the scroll regions.

However, it is quite surprising that the quantum Hall effect has been more difficult to observe in ultra-clean suspended graphene [88, 89] rather than for graphene over a SiO_2 substrate. A possible explanation is that suspended graphene tends to form scrolls at the edges, where the layer is no longer orthogonal to the magnetic field, whose effective value changes according to the curvature [90]. This motivated our theoretical study of magnetotransport in large graphene ribbons with scrolled edges [91], which shows that the effective magnetic field, i.e. its component orthogonal to the layer, oscillates and changes its sign across the scrolls, thus inducing extra non-chiral channels. Owing to the lack of spatial chirality, the scroll states easily give rise to backscattering in the presence of even moderate disorder. Note that back-scattering occurs within each scroll, while no current flows between the two opposite edges of the sample. However, due to the high quality of the samples, these extra conductive channels are not (completely) quenched and contribute to electron transport with a non-quantized conductance.

Only in the presence of strong disorder could we expect a complete suppression of the non-chiral channel contribution due to the localization of the states, and the consequent restoration of the conductance quantization. Figure 7.10(a) sketches this mechanism by indicating that non-chiral channels are present in the scrolls, while standard Landau states and edge states (at the interface with the scrolls) are present in the flat part of the ribbon. In the case of a 100 nm wide ribbon with 20 nm wide scrolls (with two turns each), the quantized conductance is increased with respect to that of the flat ribbon, due to the extra states in the scrolls; see the dashed line in figure 7.10(b). In the same panel, the continuous line corresponds to the conductance in the presence of 50 Gaussian impurities with strength in the range [−500, 500] meV and spatial range 2 nm, and of Anderson disorder with strength [−25, 25] meV. As discussed above, the conductance decreases due to the scattering within the non-chiral channels of the scrolls, and between the non-chiral channels of the scrolls and the neighbor edge channels close to the scrolls in the flat region of the ribbon. This mechanism is illustrated in figure 7.10(c), which shows the spectral transport current distribution at a representative electron energy $E = 170$ meV. The electrons are injected from the left and, for the sake of clarity, the scrolls are 'unrolled' and their borders are indicated by a dashed line. The current flows along both the top and the bottom scroll, with some backscattering due to disorder. Moreover, we can observe that electrons also flow along the top (chiral) edge channel, with a partial scattering into the non-chiral scroll channels (across the top dashed line). In contrast, the bottom edge channel is almost empty (it would be completely empty in the absence of the scrolls, since it only supports electrons moving from right to left) and only partially populated by electrons scattered from the channels of the bottom scroll (across the bottom dashed line). Note that no current flows across the inner part of the ribbon in between the two edges.

In order to obtain metrological standards of electrical resistance based on the quantum Hall effect, we thus need supported graphene of very large size. To this aim, a promising fabrication technique is chemical vapor deposition (CVD) of monolayer graphene on metals, which is then reported on insulating substrates. However, this technique entails the formation of polycrystalline graphene with grain sizes of few μm [92], whose effect has turned out to be deleterious on the Hall conductance quantization accuracy required for metrological purposes, with deviations around 1% [93]. The power-law, and non-exponential, temperature scaling of the longitudinal resistance excludes the variable range hopping regime as the main origin of the dissipative transport. Indeed, our numerical simulations [93, 94] demonstrate that extended states can develop along the grain boundaries, which constitute a net of conductive channels able to bring electrons from one edge to the other of the Hall bar, and thus to break the spatial chirality of the currents and make transport dissipative; see figure 7.11. The conductive states along the grain boundaries result from the combination of the oppositely-travelling edge states of the two grains. As a consequence, these channels are weakly chiral, in the sense that the spatial separation between them is small, and roughly corresponding to the grain boundary width. Depending on the specific grain boundary geometry and on the considered electron energy, the states can hybridize into non-chiral channels or

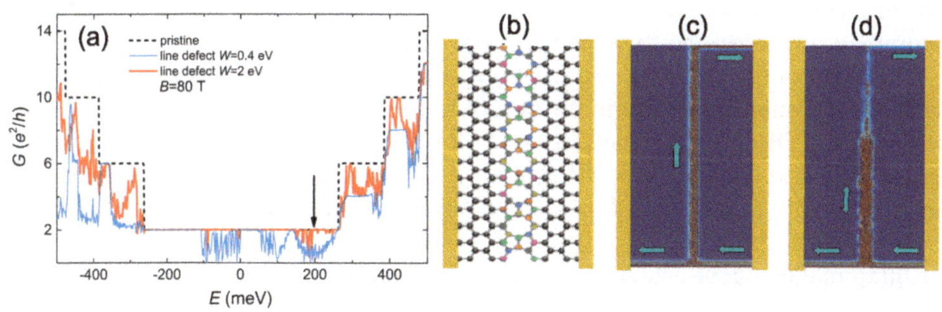

Figure 7.11. (a) Two-terminal magnetoconductance of 100 nm wide armchair ribbon in the pristine case and in the presence of a 8–5 line defect crossing the sample (represented in (b)) including a random Anderson disorder with strength 0.4 eV (blue line) and 2 eV (red line). (b) Sketch of the 8–5 line defect crossing the aGNR. The colored dots correspond to the atoms where the Anderson disorder is active. (c), (d) Spatial distribution of the electrons injected from the source contact (to the right) at 200 meV indicated by an arrow in (a), for Anderson disorder strength 0.4 eV (c) and 2 eV (d). Reprinted with permission from [93], Copyright (2014) by the American Physical Society.

locally open a gap. As a result of the weak chirality and of the narrow spatial distribution of the grain boundary states along the line defects themselves, they are extremely sensitive to disorder and prone to localization. This is shown in figure 7.11(a), where, in the presence of increasing Anderson disorder along the grain boundary (which mimics, for example, passivation by different chemical species), the conductance quantization is progressively restored. The local density of occupied states reported in figures 7.11(c) and (d) clearly illustrates that while, for a low-disordered grain boundary, electrons can flow along it and be partially back-scattered from top to bottom edge, electrons cannot reach the bottom edge and spatial chirality is preserved in the presence of stronger disorder. This mechanism explains why, in large and disordered experimental samples, the deviation from quantized conductance is found to be much smaller than in our simulations. Such a behavior has been numerically shown to be valid for different grain boundary geometries [95], and also to affect shot-noise and non-local resistance in a more realistic six terminal all bar configuration [96, 97]. A Hall resistance quantization with metrological quality can be obtained by CVD graphene directly grown over a SiC substrate [73]. Here, in addition to the higher quality of the samples, the substrate-to-graphene charge transfer [98, 99] assures the observation of the $\nu = 2$ plateau over an extremely large range of magnetic fields.

7.7 Conclusion

In this chapter, we gave an overview of recent progress in the understanding of electronic transport properties of edge disordered and chemically modified graphene nanoribbons. The case of substitutional doping (boron, nitrogen) has allowed us to pinpoint the possibility to induce some marked electron–hole transport asymmetry and mobility gaps, which could enhance the performance of graphene-based field-effect transistors. Similarly, electron–hole transport asymmetries have been found to occur for certain types of edge defects (pentagons, heptagons), but with an enhanced sensitivity to the chemical passivation degree. Finally, covalent adsorptions (sp^3 defects) were shown to dramatically affect the conducting properties of graphene materials.

The effect of disorder has also been investigated in the case of magnetotransport, which is particularly relevant in the case of the quantum Hall effect for metrological applications. Disorder, such as polycrystallinity or edge scrolling, is found to play a central role in the breaking of the spatial chirality of edge currents, and then on the formation and the quality of quantized Hall conductance plateaus. In narrow ribbons, the specific nature of disorder was shown to be strictly correlated to the resulting magnetoconductance trend, as experimentally observed.

Acknowledgments

Los Alamos National Laboratory is managed by Triad National Security, LLC, for the National Nuclear Security Administration of the U.S. Department of Energy under Contract No. 89233218CNA000001. A L-B acknowledges the support provided by the US Department of Energy, Office of Science, Basic Energy Sciences, Materials

Science and Engineering Division. J-C C acknowledges financial support from the Fédération Wallonie-Bruxelles through the Action de Recherche Concerte (ARC) on 3D nanoarchitecturing of 2D crystals (No. 16/21–077), from the European Unions Horizon 2020 researchers and innovation programme (No. 696656), and from the Fonds de la Recherche Scientifique de Belgique (F.R.S.-FNRS).

References

[1] Charlier J-C, Blase X and Roche S 2007 Electronic and transport properties of nanotubes *Rev. Mod. Phys.* **79** 677–732

[2] Neto A H C, Guinea F, Peres N M R, Novoselov K S and Geim A K 2009 The electronic properties of graphene *Rev. Mod. Phys.* **81** 109–62

[3] Cresti A, Nemec N, Biel B, Niebler G, Triozon F, Cuniberti G and Roche S 2008 Charge transport in disordered graphene-based low dimensional materials *Nano Res.* **1** 361–94

[4] Lemme M C 2010 Current status of graphene transistors *Gettering and Defect Engineering in Semiconductor Technology XIII Solid State Phenomena* vol 156 (Trans Tech Publications), pp 499–509

[5] Katsnelson M I, Novoselov K S and Geim A K 2006 Chiral tunnelling and the Klein paradox in graphene *Nat. Phys.* **2** 620–25

[6] McCann E, Kechedzhi K, Fal'ko V I, Suzuura H, Ando T and Altshuler B L 2006 Weak-localization magnetoresistance and valley symmetry in graphene *Phys. Rev. Lett.* **97** 146805

[7] Tikhonenko F V, Kozikov A A, Savchenko A K and Gorbachev R V 2009 Transition between electron localization and antilocalization in graphene *Phys. Rev. Lett.* **103** 226801

[8] Loh K P, Bao Q, Ang P K and Yang J 2010 The chemistry of graphene *J. Mater. Chem.* **20** 2277

[9] Calzolari A, Marzari N, Souza I and Nardelli M B 2004 *Ab initio* transport properties of nanostructures from maximally localized wannier functions *Phys. Rev.* B **69** 035108

[10] Lee Y-S and Marzari N 2006 Cycloaddition functionalizations to preserve or control the conductance of carbon nanotubes *Phys. Rev. Lett.* **97** 116801

[11] Lee Y-S, Nardelli M B and Marzari N 2005 Band structure and quantum conductance of nanostructures from maximally localized Wannier functions: the case of functionalized carbon nanotubes *Phys. Rev. Lett.* **95** 076804

[12] Artacho E, Sánchez-Portal D, Ordejón P, García A and Soler J M 1999 Linear-scaling ab-initio calculations for large and complex systems *Phys. Status Solidi* b **215** 809–17

[13] Ordejón P, Artacho E and Soler J M 1996 Self-consistent order-*n* density-functional calculations for very large systems *Phys. Rev.* B **53** R10441–4

[14] Demkov A A, Ortega J, Sankey O F and Grumbach M P 1995 Electronic structure approach for complex silicas *Phys. Rev.* B **52** 1618–30

[15] Fisher D S and Lee P A 1981 Relation between conductivity and transmission matrix *Phys. Rev.B* **23** 6851–68

[16] Guinea F, Tejedor C, Flores F and Louis E 1983 Effective two-dimensional Hamiltonian at surfaces *Phys. Rev.* B **28** 4397–402

[17] Lopez Sancho M P, Lopez Sancho J M and Rubio J 1984 Quick iterative scheme for the calculation of transfer matrices: application to Mo (100) *J. Phys. F: Met. Phys.* **14** 1205–15

[18] Adessi C, Roche S and Blase X 2006 Reduced backscattering in potassium-doped nanotubes: *ab initio* and semiempirical simulations *Phy. Rev.* B **73** 125414

[19] Nardelli M B 1999 Electronic transport in extended systems: application to carbon nanotubes *Phys. Rev.* B **60** 7828–33
[20] Triozon F, Lambin P and Roche S 2005 Electronic transport properties of carbon nanotube based metal/semiconductor/metal intramolecular junctions *Nanotechnology* **16** 230–33
[21] Gómez-Navarro C, De Pablo P J, Gómez-Herrero J, Biel B, Garcia-Vidal F J, Rubio A and Flores F 2005 Tuning the conductance of single-walled carbon nanotubes by ion irradiation in the Anderson localization regime *Nat. Mater.* **4** 534–39
[22] López-Bezanilla A, Triozon F, Latil S, Blase X and Roche S 2009 Effect of the chemical functionalization on charge transport in carbon nanotubes at the mesoscopic scale *Nano Lett.* **9** 940–44
[23] López-Bezanilla A, Triozon F and Roche S 2009 Chemical functionalization effects on armchair graphene nanoribbon transport *Nano Lett.* **9** 2537–41
[24] Markussen T, Rurali R, Brandbyge M and Jauho A-P 2006 Electronic transport through Si nanowires: role of bulk and surface disorder *Phys. Rev.* B **74** 245313
[25] Rocha A R, Rossi M, Fazzio A and da Silva A J R 2008 Designing real nanotube-based gas sensors *Phys. Rev. Lett.* **100** 176803
[26] Biel B, García-Vidal F J, Rubio Á and Flores F 2008 *Ab initio* study of transport properties in defected carbon nanotubes: an O(N) approach *J. Phys. Condens. Matter* **20** 294214
[27] Rod R 2008 Calling all chemists *Nat. Nanotechnol.* **3** 10–1
[28] Biel B, Triozon F, Blase X and Roche S 2009 Chemically induced mobility gaps in graphene nanoribbons: a route for upscaling device performances *Nano Lett.* **9** 2725–29
[29] Cervantes-Sodi F, Csányi G, Piscanec S and Ferrari A C 2008 Edge-functionalized and substitutionally doped graphene nanoribbons: electronic and spin properties *Phys. Rev.* B **77** 165427
[30] Martins T B, Miwa R H, da Silva A J R and Fazzio A 2007 Electronic and transport properties of boron-doped graphene nanoribbons *Phys. Rev. Lett.* **98** 196803
[31] Bronner C, Stremlau S, Gillev M, Brauße F, Haase A, Hecht S and Tegeder P 2013 Aligning the band gap of graphene nanoribbons by monomer doping *Angew. Chem. Int. Ed.* **52** 4422–25
[32] Carbonell-Sanromà E *et al* 2017 Quantum dots embedded in graphene nanoribbons by chemical substitution *Nano Lett.* **17** 50–6
[33] Cloke R R, Marangoni T, Nguyen G D, Joshi T, Rizzo D J, Bronner C, Cao T, Louie S G, Crommie M F and Fischer F R 2015 Site-specific substitutional boron doping of semiconducting armchair graphene nanoribbons *J. Am. Chem. Soc.* **137** 8872–75
[34] Kawai S, Nakatsuka S, Hatakeyama T, Pawlak R, Meier T, Tracey J, Meyer E and Foster A S 2018 Multiple heteroatom substitution to graphene nanoribbon *Sci. Adv.* **4** eaar7181
[35] Kawai S, Saito S, Osumi S, Yamaguchi S, Foster A S, Spijker P and Meyer E 2015 Atomically controlled substitutional boron-doping of graphene nanoribbons *Nat. Commun.* **6** 8098
[36] Nguyen G D *et al* 2016 Bottom-up synthesis of $n = 13$ sulfur-doped graphene nanoribbons *J. Phys. Chem.* C **120** 2684–87
[37] Senkovskiy B V *et al* 2018 Boron-doped graphene nanoribbons: electronic structure and Raman fingerprint *ACS Nano* **12** 7571–82
[38] Wang X-Y *et al* 2018 Bottom-up synthesis of heteroatom-doped chiral graphene nanoribbons *JACS* **140** 9104–107

[39] Zhang Y, Zhang Y, Li G, Lu J, Lin X, Du S, Berger R, Feng X, Müllen K and Gao H-J 2014 Direct visualization of atomically precise nitrogen-doped graphene nanoribbons *Appl. Phys. Lett.* **105** 023101

[40] Wang X, Li X, Zhang L, Yoon Y, Weber P K, Wang H, Guo J and Dai H 2009 Robust dirac point in honeycomb-structure nanoribbons with zigzag edges *Science* **324** 768–71

[41] Yan Q, Huang B, Yu J, Zheng F, Zang J, Wu J, Gu B-L, Liu F and Duan W 2007 Intrinsic current-voltage characteristics of graphene nanoribbon transistors and effect of edge doping *Nano Lett.* **7** 1469–73

[42] Avriller R, Latil S, Triozon F, Blase X and Roche S 2006 Chemical disorder strength in carbon nanotubes: magnetic tuning of quantum transport regimes *Phys. Rev.* B **74** 121406

[43] Avriller R, Roche S, Triozon F, Blase X and Latil S 2007 Low dimensional quantum transport properties of chemically disordered carbon nanotubes: from weak to strong localization regimes *Mod. Phys. Lett.* B **21** 1955–82

[44] Biel B, Blase X, Triozon F and Roche S 2009 Anomalous doping effects on charge transport in graphene nanoribbons *Phys. Rev. Lett.* **102** 096803

[45] Lherbier A, Blase X, Niquet Y-M, Triozon F and Roche S 2008 Charge transport in chemically doped 2D graphene *Phys. Rev. Lett.* **101** 036808

[46] Nakada K, Fujita M, Dresselhaus G and Dresselhaus M S 1996 Edge state in graphene ribbons: nanometer size effect and edge shape dependence *Phys. Rev.* B **54** 17954–61

[47] Barone V, Hod O and Scuseria G E 2006 Electronic structure and stability of semi-conducting graphene nanoribbons *Nano Lett.* **6** 2748–54

[48] Son Y-W, Cohen M L and Louie S G 2006 Energy gaps in graphene nanoribbons *Phys. Rev. Lett.* **97** 216803

[49] Magna A L, Deretzis I, Forte G and Pucci R 2009 Conductance distribution in doped and defected graphene nanoribbons *Phys. Rev.* B **80** 195413

[50] Areshkin D A, Gunlycke D and White C T 2007 Ballistic transport in graphene nanostrips in the presence of disorder: importance of edge effects *Nano Lett.* **7** 204–10

[51] Cresti A and Roche S 2009 Edge-disorder-dependent transport length scales in graphene nanoribbons: from Klein defects to the superlattice limit *Phys. Rev.* B **79** 233404

[52] Evaldsson M, Zozoulenko I V, Xu H and Heinzel T 2008 Edge-disorder-induced anderson localization and conduction gap in graphene nanoribbons *Phys. Rev.* B **78** 161407

[53] Querlioz D, Apertet Y, Valentin A, Huet K, Bournel A, Galdin-Retailleau S and Dollfus P 2008 Suppression of the orientation effects on bandgap in graphene nanoribbons in the presence of edge disorder *Appl. Phys. Lett.* **92** 042108

[54] Lherbier A, Biel B, Niquet Y-M and Roche S 2008 Transport length scales in disordered graphene-based materials: strong localization regimes and dimensionality effects *Phys. Rev. Lett.* **100** 036803

[55] Dubois S M-M, Lopez-Bezanilla A, Cresti A, Triozon F, Biel B, Charlier J-C and Roche S 2010 Quantum transport in graphene nanoribbons: effects of edge reconstruction and chemical reactivity *ACS Nano* **4** 1971–76

[56] Koskinen P, Malola S and Häkkinen H 2008 Self-passivating edge reconstructions of graphene *Phys. Rev. Lett.* **101** 115502

[57] Wang S, Talirz L, Pignedoli C A, Feng X, Müllen K, Fasel R and Ruffieux P 2016 Giant edge state splitting at atomically precise graphene zigzag edges *Nat. Commun.* **7** 11507

[58] Ruffieux P *et al* 2016 On-surface synthesis of graphene nanoribbons with zigzag edge topology *Nature* **531** 489–92

[59] Salemi L, Lherbier A and Charlier J-C 2018 Spin-dependent properties in zigzag graphene nanoribbons with phenyl-edge defects *Phys. Rev.* B **98** 214204
[60] Boukhvalov D W and Katsnelson M I 2008 Chemical functionalization of graphene with defects *Nano Lett.* **8** 4373–79
[61] Zhao J, Park H, Han J and Lu J P 2004 Electronic properties of carbon nanotubes with covalent sidewall functionalization *J. Phys. Chem.* B **108** 4227–30
[62] Boukhvalov D W and Katsnelson M I 2008 Modeling of graphite oxide *J. Am. Chem. Soc.* **130** 10697–701
[63] Yazyev O V and Helm L 2007 Defect-induced magnetism in graphene *Phys. Rev.* B **75** 125408
[64] Sofo J O, Chaudhari A S and Barber G D 2007 Graphene: a two-dimensional hydrocarbon *Phys. Rev.* B **75** 153401
[65] Roman T, Diño W A, Nakanishi H, Kasai H, Sugimoto T and Tange K 2006 Realizing a carbon-based hydrogen storage material *Japan. J. Appl. Phys.* **45** 1765–67
[66] Boukhvalov D W and Katsnelson M I 2009 Chemical functionalization of graphene *J. Phys.: Condens. Matter* **21** 344205
[67] Boukhvalov D W and Katsnelson M I 2008 Tuning the gap in bilayer graphene using chemical functionalization: density functional calculations *Phys. Rev.* B **78** 085413
[68] Jiang D, Sumpter B G and Dai S 2006 How do aryl groups attach to a graphene sheet? *J. Phys. Chem.* B **110** 23628–32
[69] Farmer D B, Golizadeh-Mojarad R, Perebeinos V, Lin Y-M, Tulevski G S, Tsang J C and Avouris P 2009 Chemical doping and electron-hole conduction asymmetry in graphene devices *Nano Lett.* **9** 388–92
[70] Lopez-Bezanilla A 2014 Uniform quantum transport properties across different chemical decorations in bilayer graphene *J Phys. Chem.* C **118** 29467–72
[71] Zheng Y and Ando T 2002 Hall conductivity of a two-dimensional graphite system *Phys. Rev.* B **65** 245420
[72] Zhang Y, Tan Y-W, Stormer H L and Kim P 2005 Experimental observation of the quantum Hall effect and Berry's phase in graphene *Nature* **438** 201–04
[73] Lafont F *et al* 2015 Quantum Hall resistance standards from graphene grown by chemical vapour deposition on silicon carbide *Nat. Commun.* **6** 6806
[74] Klitzing K V, Dorda G and Pepper M 1980 New method for high-accuracy determination of the fine-structure constant based on quantized Hall resistance *Phys. Rev. Lett.* **45** 494–97
[75] Thouless D J, Kohmoto M, Nightingale M P and den Nijs M 1982 Quantized Hall conductance in a two-dimensional periodic potential *Phys. Rev. Lett.* **49** 405–08
[76] Peierls R 1933 Zur theorie des diamagnetismus von leitungselektronen *Z. Phys.* **80** 763–91
[77] Huang Y C, Chang C P and Lin M F 2007 Magnetic and quantum confinement effects on electronic and optical properties of graphene ribbons *Nanotechnology* **18** 495401
[78] Wakabayashi K, Fujita M, Ajiki H and Sigrist M 1999 Electronic and magnetic properties of nanographite ribbons *Phys. Rev.* B **59** 8271–82
[79] Poumirol J-M, Cresti A, Roche S, Escoffier W, Goiran M, Wang X, Li X, Dai H and Raquet B 2010 Edge magnetotransport fingerprints in disordered graphene nanoribbons *Phys. Rev.* B **82** 041413
[80] Li X, Wang X, Zhang L, Lee S and Dai H 2008 Chemically derived, ultrasmooth graphene nanoribbon semiconductors *Science* **319** 1229–32

[81] Shen H, Cresti A, Escoffier W, Shi Y, Wang X and Raquet B 2015 Peculiar magnetotransport features of ultranarrow graphene nanoribbons under high magnetic field *ACS Nano* **10** 1853–58

[82] Fujita M, Wakabayashi K, Nakada K and Kusakabe K 1996 Peculiar localized state at zigzag graphite edge *J. Phys. Soc. Jpn.* **65** 1920–23

[83] Ribeiro R, Poumirol J-M, Cresti A, Escoffier W, Goiran M, Broto J-M, Roche S and Raquet B 2011 Unveiling the magnetic structure of graphene nanoribbons *Phys. Rev. Lett.* **107** 086601

[84] Pereira A L C and Schulz P A 2008 Additional levels between Landau bands due to vacancies in graphene: towards defect engineering *Phys. Rev.* B **78** 125402

[85] Leconte N, Ortmann F, Cresti A, Charlier J-C and Roche S 2014 Quantum transport in chemically functionalized graphene at high magnetic field: defect-induced critical states and breakdown of electron-hole symmetry *2D Mater.* **1** 021001

[86] Leconte N, Ortmann F, Cresti A and Roche S 2016 Unconventional features in the quantum Hall regime of disordered graphene: percolating impurity states and Hall conductance quantization *Phys. Rev.* B **93** 115404

[87] Petrović M D and Peeters F M 2016 Quantum transport in graphene Hall bars: effects of vacancy disorder *Phys. Rev.* B **94** 235413

[88] Bolotin K I, Ghahari F, Shulman M D, Stormer H L and Kim P 2009 Observation of the fractional quantum Hall effect in graphene *Nature* **462** 196–99

[89] Du X, Skachko I, Duerr F, Luican A and Andrei E Y 2009 Fractional quantum Hall effect and insulating phase of Dirac electrons in graphene *Nature* **462** 192–95

[90] Ferrari G, Bertoni A, Goldoni G and Molinari E 2008 Cylindrical two-dimensional electron gas in a transverse magnetic field *Phys. Rev.* B **78** 115326

[91] Cresti A, Fogler M M, Guinea F, Neto A H C and Roche S 2012 Quenching of the quantum Hall effect in graphene with scrolled edges *Phys. Rev. Lett.* **108** 166602

[92] Kim K, Lee Z, Regan W, Kisielowski C, Crommie M F and Zettl A 2011 Grain boundary mapping in polycrystalline graphene *ACS Nano* **5** 2142–46

[93] Lafont F *et al* 2014 Anomalous dissipation mechanism and Hall quantization limit in polycrystalline graphene grown by chemical vapor deposition *Phys. Rev.* B **90** 115422

[94] Cummings A W, Cresti A and Roche S 2014 Quantum Hall effect in polycrystalline graphene: the role of grain boundaries *Phys. Rev.* B **90** 161401

[95] Bergvall A, Carlsson J M and Löfwander T 2015 Influence of [0001] tilt grain boundaries on the destruction of the quantum Hall effect in graphene *Phys. Rev.* B **91** 245425

[96] Lago V D and Foa Torres L E F 2015 Line defects and quantum Hall plateaus in graphene *J. Phys.: Condens. Matter* **27** 145303

[97] Phillips M and Mele E J 2017 Charge and spin transport on graphene grain boundaries in a quantizing magnetic field *Phys. Rev.* B **96** 041403

[98] Kopylov S, Tzalenchuk A, Kubatkin S and Fal'ko V I 2010 Charge transfer between epitaxial graphene and silicon carbide *Appl. Phys. Lett.* **97** 112109

[99] Yang M *et al* 2016 Puddle-induced resistance oscillations in the breakdown of the graphene quantum Hall effect *Phys. Rev. Lett.* **117** 237702

Lightning Source UK Ltd.
Milton Keynes UK
UKHW052048140920
369737UK00003BA/42

9 780750 316996